Studies in Computational Intelligence

Volume 766

Series editor

Janusz Kacprzyk, Polish Academy of Sciences, Warsaw, Poland
e-mail: kacprzyk@ibspan.waw.pl

The series "Studies in Computational Intelligence" (SCI) publishes new developments and advances in the various areas of computational intelligence—quickly and with a high quality. The intent is to cover the theory, applications, and design methods of computational intelligence, as embedded in the fields of engineering, computer science, physics and life sciences, as well as the methodologies behind them. The series contains monographs, lecture notes and edited volumes in computational intelligence spanning the areas of neural networks, connectionist systems, genetic algorithms, evolutionary computation, artificial intelligence, cellular automata, self-organizing systems, soft computing, fuzzy systems, and hybrid intelligent systems. Of particular value to both the contributors and the readership are the short publication timeframe and the world-wide distribution, which enable both wide and rapid dissemination of research output.

More information about this series at http://www.springer.com/series/7092

Iwona Skalna

Parametric Interval Algebraic Systems

Springer

Iwona Skalna
Department of Applied Computer Science,
Faculty of Management
AGH University of Science
and Technology
Kraków
Poland

ISSN 1860-949X ISSN 1860-9503 (electronic)
Studies in Computational Intelligence
ISBN 978-3-319-89287-0 ISBN 978-3-319-75187-0 (eBook)
https://doi.org/10.1007/978-3-319-75187-0

Printed on acid-free paper

This Springer imprint is published by the registered company Springer International Publishing AG part of Springer Nature
The registered company address is: Gewerbestrasse 11, 6330 Cham, Switzerland

Contents

1 Interval Arithmetic 1
1.1 Intervals 3
1.2 Basic Properties of Interval Arithmetic 6
1.3 Range Bounding 7
1.4 Rounded Interval Arithmetic 20
1.4.1 Setting Rounding Mode 23

2 Alternative Arithmetic 25
2.1 Affine Arithmetic 25
2.1.1 Arithmetic Operations on Affine Forms 28
2.1.2 Affine Approximations of Basic Univariate Nonlinear Functions 30
2.2 Revised Affine Arithmetic 34
2.2.1 Multiplication 35
2.2.2 Reciprocal 46
2.2.3 Division 49
2.2.4 Rounded Revised Affine Arithmetic 50

3 Interval and Parametric Interval Matrices 51
3.1 Interval Matrix 51
3.2 Parametric Interval Matrix 57
3.3 Affine Transformation 58
3.4 NP-Hard Problems Related to Parametric Interval Matrices 59
3.5 Regularity 60
3.6 Strong Regularity 64
3.6.1 Pre-conditioning with Mid-Point Inverse 64
3.6.2 Pre-conditioning with Arbitrary Non-singular Matrix 68
3.7 Radius of Regularity 72
3.8 Radius of Strong Regularity 73
3.9 Positive (Semi-)Definiteness 74

3.10 Hurwitz Stability 78
3.11 Schur Stability 79
3.12 Radius of Stability 81
3.13 Inverse Stability 81

4 Linear Systems 85
4.1 Interval Linear Systems 85
4.2 Parametric Interval Linear Systems 87
4.3 Shape of (United) Parametric Solution Set 88
4.4 Affine Transformation of System of Linear Equations 94
4.5 Complex Parametric Interval Linear Systems 96
4.6 Over- and Under-Determined Parametric Interval Linear Systems 97

5 Methods for Solving Parametric Interval Linear Systems 99
5.1 Methods for Approximate Solutions 101
5.1.1 Modified Rump's Fixed-Point Iteration 101
5.1.2 Parametric Gauss–Seidel Iteration 105
5.1.3 Direct Method 107
5.1.4 Generalized Parametric Bauer–Skeel Method 110
5.1.5 Generalized Parametric Hansen–Bliek–Rohn Method 113
5.1.6 Monotonicity Approach 118
5.2 Methods for Computing Parametric Solution 121
5.2.1 Parametric Methods 121
5.2.2 Interval-Affine Gauss–Seidel Iteration 124
5.2.3 Convergence of Iterative Methods 128
5.2.4 Generalized Expansion Method 133
5.3 Methods for Interval Hull Solution 136
5.3.1 Parametric Interval Global Optimization 137
5.3.2 Evolutionary Strategy 140
5.4 Comparison of Enclosure Methods 143
5.4.1 Parametric Systems with Affine-Linear Dependencies 144
5.4.2 Parametric Systems with Nonlinear Dependencies 149
5.4.3 Practical Applications 153

6 Parametric Interval Linear Programming Problem 165
6.1 Iterative Method 166

Summary 173

References 177

Index 189

Notations

Notation used in this book follows the notation of IEEE Std 1788-2015, Standard for Interval Arithmetic [52]. Real scalars and vectors are denoted by small letters (e.g., $a, b, c, x, y, z, \ldots$). Real matrices are denoted by capital letters mainly from the beginning of the alphabet (e.g., A, B, C, ...). The corresponding interval quantities are distinguished by using bold letters.

$\boldsymbol{A} = (\boldsymbol{A}_{ij}) = [\underline{A}, \overline{A}]$	Interval matrix, p. 51
$\underline{A} = \inf(\boldsymbol{A})$	Infimum of interval matrix $\boldsymbol{A}$ (left endpoint matrix), p. 51
$A = \sup(\boldsymbol{A})$	Supremum of interval matrix $\boldsymbol{A}$ (right endpoint matrix), p. 51
A	Element of interval matrix $\boldsymbol{A}$, p. 52
A_{*i}	i-th column of matrix A, p. 82
$A_{j:}$	j-th row of matrix A, p. 82
$\boldsymbol{A}^{-1}$	Inverse of interval matrix $\boldsymbol{A}$, p. 53
$A(p)$	Real parametric matrix, p. 57
$A(\boldsymbol{p})$	Parametric interval matrix, p. 57
$\mathrm{b}_I(P)$	Bernstein coefficient of polynomial P, p. 15
$B_{N,I}(x)$	I-th Bernstein polynomial of degree N, p. 15
$\boldsymbol{C}(e)$	Matrix of revised affine forms, p. 58
$d_H(\boldsymbol{x}, \boldsymbol{y})$	Hausdorff distance between intervals $\boldsymbol{x}$ and $\boldsymbol{y}$, p. 5
$\mathrm{Dom}(f)$	Domain of function f, p. 8
$\boldsymbol{f}$	Interval extension of function f, p. 7
f	Expression producing function f by evaluation, p. 7
f^a	Affine approximation of nonlinear function f, p. 29
$\boldsymbol{f}_C(\boldsymbol{x})$	Centered form, p. 11
$\boldsymbol{f}_M(\boldsymbol{x})$	Mean value form, p. 11
$\boldsymbol{f}_G(\boldsymbol{x})$	Generalized centered form, p. 12
$\boldsymbol{f}_B(\boldsymbol{x})$	Bicentered form, p. 12
$f \searrow$	Decreasing function, p. 17
$f \nearrow$	Increasing function, p. 17
$\mathbb{F}$	Set of machine numbers, p. 20
fl	Rounding to nearest, p. 23

hull (S)	(Interval) hull of set $S \subseteq \mathbb{R}^n$, p. 5
$\mathbb{IF}$	Set of machine representable intervals, p. 20
iff	If and only if, p. 55
$\inf(\boldsymbol{A}) = \underline{A}$	Infimum of interval matrix $\boldsymbol{A}$ (left endpoint matrix), p. 51
$\inf(\boldsymbol{x}) = \underline{x}$	Lower bound (left endpoint) of interval $\boldsymbol{x}$, p. 3
$\mathbb{R}$	Set of real numbers, p. 3
$\overline{\mathbb{R}}$	Set of extended real numbers $\mathbb{R} \cup \{-\infty, +\infty\}$, p. 3
$\overline{\mathbb{IR}}$	Set of closed real intervals, including unbounded intervals and the empty set, p. 3
$\mathbb{IR}^*$	Set of improper intervals, p. 18
$\mathbb{IR}^n$	Set of interval vectors (boxes), p. 51
$\mathbb{IR}^{m \times n}$	Set of $m \times n$ interval matrices, p. 51
$\mathbb{KR} = \mathbb{IR} \cup \mathbb{IR}^*$	Set of proper and improper intervals, p. 18
mid$(\boldsymbol{A})$=A^c	Midpoint of interval matrix $\boldsymbol{A}$, p. 51
$\boldsymbol{p}(\rho)$	Interval parameter with radius scaled by ρ, p. 170
$Pr(\boldsymbol{p}(\rho))$	Interval analysis problem defined for $\boldsymbol{p}(\rho)$, p. 170
$r(A(\boldsymbol{p}))$	Radius of regularity of parametric interval matrix, pp. 59, 72
rad$(\boldsymbol{A})$ = A^Δ	Radius of interval matrix $\boldsymbol{A}$, p. 51
$r_a(M)$	Radius of applicability of method M, p. 170
Rge $(f \mid S)$	Range of function f over set $S \subseteq \mathbb{R}^n$, p. 7
$r_I(\boldsymbol{A})$	Radius of regularity of interval matrix $\boldsymbol{A}$, pp. 59, 73
$r^*(A(\boldsymbol{p}))$	Radius of strong regularity of parametric interval matrix $A(\boldsymbol{p})$, p. 73
$s(A(\boldsymbol{p}))$	Radius of stability of parametric interval matrix, pp. 59, 81
$S(\boldsymbol{A}, \boldsymbol{b})$	United solution set of interval linear system $\boldsymbol{A}x = \boldsymbol{b}$, p. 86
$S(\boldsymbol{A}, \boldsymbol{b})_{ctr}$	Controllable solution set of interval linear system $\boldsymbol{A}x = \boldsymbol{b}$, p. 86
$S(\boldsymbol{A}, \boldsymbol{b})_{tol}$	Tolerable solution set of interval linear system $\boldsymbol{A}x = \boldsymbol{b}$, p. 86
$A(\boldsymbol{p})$	Parametric (united) solution set, p. 88
$S_{AE}(\boldsymbol{p})$	Parametric AE-solution set, p. 88
$S_{ctr}(\boldsymbol{p})$	Parametric controllable solution set, p. 88
$S_{tol}(\boldsymbol{p})$	Parametric tolerable solution set, p. 88
$s_f[\boldsymbol{x}, z]$	Interval slope, p. 12
sgn(x)	Sign of x, p. 40
sup $(\boldsymbol{A})$ = $\underline{A}$	Supremum of interval matrix $\boldsymbol{A}$ (right endpoint matrix), p. 51
$\sup(\boldsymbol{x}) = \overline{x}$	Upper bound (right endpoint) of interval $\boldsymbol{x}$, p. 3
$T_y = \mathrm{diag}(y)$	Square matrix with vector y on the diagonal, p. 73
$\boldsymbol{U}$	Closed interval $[-1, +1]$, p. 26
wid(x)	Width of interval $\boldsymbol{x}$, p. 4
$\boldsymbol{x}^{OI}$	Outer interval solution, p. 99
$\boldsymbol{x}^{IH}$	Interval hull solution, p. 99
$\boldsymbol{x}^{IEH}$	Inner estimate of the hull solution, p. 99
$\boldsymbol{x} = [\underline{x}, \overline{x}]$	(Mathematical) closed interval of real numbers, p. 3
$\boldsymbol{x}(p)$	Parametric solution (p-solution), p. 100

$\hat{x}$	Affine form, revised affine form, pp. 26,34
$\hat{\boldsymbol{x}}$, $[\hat{x}]$	Range of (revised) affine form $\hat{x}$, p. 30
$\underline{x} = \inf(\boldsymbol{x})$	Lower bound (left endpoint) of interval $\boldsymbol{x}$, p. 3
$\bar{x} = \sup(\boldsymbol{x})$	Upper bound (right endpoint) of interval $\boldsymbol{x}$, p. 3
$A^c = \mathrm{mid}(\boldsymbol{A})$	Midpoint of interval matrix $\boldsymbol{A}$, p. 51
$A^\Delta = \mathrm{rad}(\boldsymbol{A})$	Radius of interval matrix $\boldsymbol{A}$, p. 51
$x^c = (\bar{x} + \underline{x})/2$	Midpoint of interval $\boldsymbol{x}$, p. 3
$\mathrm{q}(\boldsymbol{x}, \boldsymbol{y})$	Distance between vectors $\boldsymbol{x}$ and $\boldsymbol{y}$, p. 166
$x^\Delta = (\bar{x} - \underline{x})/2$	Radius of interval $\boldsymbol{x}$, p. 3
λ	Eigenvalue of a real matrix, p. 54
σ	Singular value of a real matrix, p. 63
$\rho(A)$	Spectral radius of matrix A, p. 54
$\rho_0(A)$	Real maximal magnitude eigenvalue of matrix A, p. 72
φ	Elementary function, p. 22
$\varphi^\diamond$	Outward rounded value of φ, p. 22
ω	Order or comparison relation, p. 6
Ω	Set of elementary operations, p. 4
$\langle \boldsymbol{A} \rangle$	Comparison matrix of a square interval matrix $\boldsymbol{A}$, p. 51
$[x]$, $[a, b]$	(Mathematical) interval of real numbers, p. 3
$\lvert \boldsymbol{x} \rvert$	Magnitude of interval $\boldsymbol{x}$, p. 4
$\langle \boldsymbol{x} \rangle$	Magnitude of interval $\boldsymbol{x}$, p. 4
$\langle \hat{x}, \hat{y} \rangle$	Joint range of two affine forms $\hat{x}$ and $\hat{y}$, p. 27
$\langle \hat{x}, \hat{y} \rangle^+$	nonnegative quadrant of joint range $\langle \hat{x}, \hat{y} \rangle$, p. 27
$\lVert A \rVert_u$	Scaled maximum norm of square matrix A, p. 53
$\lVert x \rVert_1$	l_1-norm of vector x, p. 27
$\lVert x \rVert_u$	Scaled maximum norm of vector x, p. 53
∂S	Border of set S, p. 40
∇^2	Hessian matrix, p. 40
$\bullet$	Elementary operation, p. 5
$\Delta(x)$	Upward rounding, p. 22
$\nabla(x)$	Downward rounding, p. 22
$\diamond(x)$	Optimal outward rounding, p. 22
$\odot$	Outward rounded value of elementary operation, p. 22
$\leqslant$-isotone	Monotone increasing
$\leqslant$-antitone	Monotone decreasing

List of Figures

Fig. 1.1 Graph of function f described by expression $\mathsf{f}(x) = 1/(1 - x + x^2)$ (left); enclosures for Rge $(f \mid [0, t])$, $t = 0.1, 0.2, 0.3, 0.4$ (right) 8
Fig. 1.2 Graph of function defined by arithmetic expression $\mathsf{f}(x) = 1 - 5x + 1/3x^3$ with exact range (left); exact range, naive bounds and bounds obtained by using bicentered form with slope (right) 20
Fig. 2.1 Overestimation of IA estimates: graph of function $g(x)$ on interval $x \in [1, 2]$ (black solid line) and values of $\boldsymbol{g}(x)$ for 16 consecutive equal subintervals (top left); graph of function $h(x) = g^3(x)$ and values of $\boldsymbol{h}(x)$ for same data (top right); same computation with affine arithmetic: $\boldsymbol{g}(x)$ (bottom left), $\boldsymbol{h}(x)$ (bottom right) 26
Fig. 2.2 Joint range of affine forms $\hat{x} = 1 - \varepsilon_1 + 3\varepsilon_2 - 5\varepsilon_3 - \varepsilon_4 - 3\varepsilon_5$, $\hat{y} = 6\varepsilon_1 - 3\varepsilon_3 - 4\varepsilon_5$ (light gray region) and box $[\hat{x}] \times [\hat{y}]$ (dark gray region); points (black dots) obtained for $\varepsilon_1, \ldots, \varepsilon_5 \in \{-1, 1\}$ (left); central point (right) 28
Fig. 2.3 Geometrical illustration of minimum range approximation of univariate function f over interval $[a, b]$; $f'' > 0, f' < 0$ (left); $f'' < 0, f' < 0$ (right). 31
Fig. 2.4 Geometric illustration of affine-linear approximation of non-monotonic smooth function f over interval $[a, b]$ 32
Fig. 2.5 Graph of function $f(x, y) = x \cdot y$ with trivial affine approximation (left) and Chebyshev minimum-error affine approximation (right). 39
Fig. 2.6 Set $\langle \hat{u}, \hat{v} \rangle^+$ (gray region) with northwest boundary (black arrows), level sets of objective function (2.28) (grey lines), right-most vertex w_1 and point $w_{\max}$ at which maximum $R_{\max}$ is attained 39

Fig. 2.7 Symmetric medium degree polynomial: $47.6 - 220.8x + 476.8x^2 - 512.0x^3 + 256.0x^4 - 220.8y + 512.0xy - 512.0x^2y + 476.8y^2 - 512.0xy^2 + 512.0x^2y^2 - 512.0y^3 + 256.0y^4$: plot obtained using multiplication (2.27) (left); plot obtained using trivial multiplication (2.21) (right) 45
Fig. 2.8 Asymmetric medium degree polynomial: $((55.0/256.0) - x + 2.0x^2 - 2.0x^3 + x^4 - (55.0/64.0)y + 2.0xy - 2.0x^2y + (119.0/64.0)y^2 - 2.0xy^2 + 2.0x^2y^2 - 2.0y^3 + y^4)$: plot obtained using the multiplication (2.27) (left); plot obtained using trivial multiplication (2.21) (right). 45
Fig. 2.9 Geometrical illustration of Chebyshev minimum-error (left) and min-range (right) approximations of reciprocal function. 48
Fig. 4.1 Parametric solution set of system (4.21) depicted as set of plane curves depending on parameter p_1, $\delta \approx 0.03$ (top left) and as set of plane curves depending on parameter p_2, $\delta \approx 0.03$ (top right); parametric solution set with part of corresponding non-parametric solution set (bottom left) and with entire non-parametric solution set (bottom right) 92
Fig. 4.2 Parametric solution set of system (4.23) depicted as set of plane curves depending on parameter p_1, $\delta \approx 0.03$ (left) and as set of plane curves depending on parameter p_2, $\delta \approx 0.07$ (right) 93
Fig. 4.3 Parametric 3D solution set of system (4.24) shown from different perspectives (top left, top right); number of sample points is 50 for both p_1 and p_2; non-parametric solution set (bottom left, bottom right) with parametric solution set (dark region in bottom left plot) 93
Fig. 4.4 Solution set of system (4.26) (black solid line) and solution set of corresponding system with affine-linear dependencies (set of plane curves depending on parameter p_1, $\delta \approx 0.06$ (left), and set of plane curves depending on parameter p_2, $\delta \approx 0.06$ (right); rectangles represent interval hulls of respective solution sets 95
Fig. 4.5 Solution set of system (4.28) (left) and solution set of corresponding system with affine-linear dependencies together with original solution set (right); number of sample points is 50 for both ε_1 and ε_2. 95
Fig. 5.1 Comparison of $\rho(|\boldsymbol{G}|)$ and $\rho(|\boldsymbol{G}'|)$ for Example 5.4 103
Fig. 5.2 Solution set of system from Example 5.17 (left); solution set (black region) together with DM, PBS, PBSRC and PHBR bounds (right) 116
Fig. 5.3 Projection of parametric solution set from Example 5.34 on v_1v_2-plane ($\delta = 0.05$ [left], $\delta = 0.1$ [right]), and OI solution obtained using Expansion method (dashed line), OI solution

obtained using GEM method (solid line), and IEH solution obtained using GEM method (dotted line) 136
Fig. 5.4 Boxes processed while computing upper bound for $\boldsymbol{x}_1^{OI}$ in Example 5.40. Boxes processed by PIGO without monotonicity test (Rule A [top left], Rule B [top right]) and with monotonicity test (Rule A [bottom left], Rule B [bottom right]) 140
Fig. 5.5 Solution set of system from Example 5.43 (left); solution set with hull solution and NP and PHBRC enclosures (right).... 146
Fig. 5.6 Solution set of system from Example 5.47 depicted as set of plane curves depending on parameter p_3 (left) and as set of plane curves depending on parameter p_2 (right); rectangles represent inner and outer bounds obtained using IAGSI method 150
Fig. 5.7 One-bay structural steel frame (cf. [30, 178]) 155
Fig. 5.8 Four-bay two-floor truss structure......................... 158
Fig. 5.9 Linear electrical circuit with five nodes and eleven branches 159
Fig. 6.1 Feasible set defined by the constraint (6.17) and points at which minimum and maximum values are attained; case A (left), case B (right)............................ 170

List of Tables

Table 1.1 Comparison of selected range bounding methods for Example 1.17 20
Table 1.2 Rounding flags settings 24
Table 2.1 Results for Example 2.14: number of iterations (#Itrs) and number of boxes left in list L (#Bxs). 43
Table 2.2 Results for Example 2.14: CPU times (in seconds) and ratio of (2.27) time to the best time. 44
Table 2.3 Comparison of multiplication methods for symmetric medium degree polynomial (2.34) 45
Table 2.4 Comparison of multiplication methods for asymmetric medium degree polynomial (2.35) 46
Table 5.1 Comparison of results obtained using DM, PBS, PBSRC and PHBR methods for Example 5.17 115
Table 5.2 Comparison of PHBR and PHBRC enclosures for Example 5.19 116
Table 5.3 Comparison of parametric methods for Example 5.23 124
Table 5.4 Comparison of result obtained using EM and GEM methods for Example 5.34 ($\delta = 0.1$) 136
Table 5.5 Comparison of performance of different variants of PIGO for Example 5.40 140
Table 5.6 Result of evolutionary strategy for Example 5.40: average result (ES ave.) from 100 runs; standard deviations for each endpoint (std.dev. lb, std.dev.ub); and result of parametric interval global optimization 143
Table 5.7 Comparison of enclosures for Example 5.42 144
Table 5.8 Comparison of outer interval enclosures for Example 5.43: outer interval solution $\boldsymbol{x}^{OI}$ and average overestimation $O_w(\boldsymbol{x}^{OI}, \boldsymbol{x}^H)$ of hull solution $\boldsymbol{x}^H$ 145

Table 5.9 Sharpness of outer enclosures for Example 5.44; minimal and maximal values of overestimation measured based on inner estimate of hull solution . . . 146
Table 5.10 CPU times (in seconds) for Example 5.44; number of iterations taken by iterative methods is given in parentheses . . . 147
Table 5.11 Sharpness of outer enclosures for Example 5.45 . . . 148
Table 5.12 Comparison of computational times and number of iterations (in parentheses) for Example 5.45 . . . 148
Table 5.13 Comparison of enclosures for Example 5.46 . . . 149
Table 5.14 Sharpness of enclosures for Example 5.47 . . . 149
Table 5.15 Inner and outer bounds of IAGSI method for Example 5.47 . . . 150
Table 5.16 Sharpness of enclosures for Example 5.48 . . . 150
Table 5.17 Comparison of interval bounds of IAGSI method and bounds reported in [46] for Example 5.48 . . . 151
Table 5.18 Sharpness of outer enclosures for Example 5.49 . . . 151
Table 5.19 Comparison of computational times and number of iterations (in parentheses) for Example 5.49 . . . 151
Table 5.20 Inner and outer bounds from IAGSI method for Example 5.49 . . . 152
Table 5.21 Sharpness of outer enclosures for Example 5.50 . . . 153
Table 5.22 Comparison of computational times and number of iterations (in parentheses) for Example 5.50 . . . 153
Table 5.23 Inner and outer bounds from IAGSI method for Example 5.50 . . . 154
Table 5.24 Parameters of one-bay structural steel frame: nominal values and worst case uncertainties . . . 156
Table 5.25 Comparison of interval enclosures and computational times (last row) for one-bay steel frame . . . 156
Table 5.26 Hull solution and results of IAGSI method for one-bay steel frame example; results are rounded outwardly to 10 digits accuracy . . . 157
Table 5.27 Sharpness of outer enclosures for Example 5.52 . . . 158
Table 5.28 Comparison of computational times and number of iterations (in parentheses) for Example 5.52 . . . 158
Table 5.29 Sharpness of outer enclosures for Example 5.53 . . . 160
Table 5.30 Comparison of computational times for Example 5.53 . . . 160
Table 5.31 Comparison of results of IAGSI method and results from [182] for Example 5.53 for 10% tolerance; hull solution is given in last column . . . 161

Table 5.32 Comparison of enclosures for Example 5.54 162
Table 5.33 Comparison of computational times and number of iterations (in parentheses) for Example 5.54 162
Table 5.34 Results of the IAGSI method for Example 5.54. 163
Table 6.1 Asymptotic time complexity of methods MPILP.V1, MPILP.V2, MPILP.V3: M - number of iterations of the M2 method, n - size of PILP problem, K - number of parameters, κ - number of iterations of M1 method, g - number of generations in P2 procedure 169
Table 6.2 Hull solution to PILP problem 6.2 obtained using MPILP.V1, MPILP.V2 and MPILP.V3 methods, $\rho = 0.1$ 170
Table 6.3 Radius of applicability of the MPILP.V1, MPILP.V2 and MPILP.V3 methods. 171
Table 6.4 Data on the MPILP.V4 method, $\rho = 0.269$ 171
Table 6.5 Data on the MPILP.V4 method, $\rho = r_a(\text{MPILP.V4}) = 0.333$. 172
Table 6.6 Data on the MPILP.V4 method, $\rho = 0.595$, $\tau = 0.74$ 172

Introduction

Computer science (CS) has many areas, including *numerical analysis* (NA) that creates, analyzes, and implements algorithms. One of the main goals of numerical analysis is to develop general-purpose algorithmic tools and paradigms for obtaining approximate but accurate solutions to hard problem, where quickly finding real solutions is not likely. Another goal is to try and improve upon existing algorithms so as to reduce the amount of errors generated by computer calculation. Numerical analysis is mathematical and abstract in spirit—the implementation of algorithms and data structures is based on mathematical theory and theorems in linear algebra, combinatorics, and analysis—but it derives its motivation from practical and everyday computation. Its aim is to understand the nature of computation and, as a consequence of this understanding, provide more efficient methodologies. Much of numerical analysis is close to engineering in the sense that its creations must work in the real world.

One of the fundamental problems of numerical analysis is the problem of solving *linear algebraic systems*. This problem, being of great importance in various fields of science and engineering, is the main topic of this book. The reason for the widespread popularity and wide applications of linear systems in various fields is that linear system theories and techniques are relatively mature, simple, and universal. Moreover, linear systems arise in a natural way in many of the problems of numerical analysis, a reflection of the approximation of mathematical problems using linearization, which is a very helpful technique when making a mathematical model or a computer simulation of a relatively complex system (as an example, numerical methods for solving partial differential equations often lead to very large "sparse" linear systems).

Not surprisingly, a lot of effort has been put in developing methods, algorithms, and software for solving systems of linear equations. However, all numerical calculations are carried out using finite precision *computer arithmetic*, usually in a framework of floating-point representation of numbers. Because of *rounding errors* that are inherent in finite machine arithmetic and *truncation errors*, numerical solutions are not guaranteed to be correct and may be afflicted by uncertainty. For ill-conditioned problems, the obtained solutions can be far from the exact ones. Furthermore, the parameters of a modeled system cannot be stored on a computer with infinite precision, so we lose in accuracy even before starting a computation. The superposition of data and computing errors can lead to catastrophic consequences (cf. [53, 234, 284]). So, there is a need for tools that will be able to track rounding and truncation errors and produce reliable results. A major breakthrough toward developing such tools was made in the late 1950s along with the pioneering work of Moore and Yang on *interval arithmetic* (IA) [151]. Moore and Yang devised interval arithmetic in order to perform a concomitant analysis of rounding errors in a digital computation. In their system, the computation was performed on intervals,[1] which enclosed a range of real numbers between two machine numbers, and a procedure known as *outward rounding* was used to guarantee that the obtained solutions are rigorous (i.e., they always, with absolute certainty, contain the exact numerical solutions sought), despite the rounding errors. Other works devoted to interval computation appeared independently in Poland, Japan, Russia, and USA (see, e.g., [94, 266, 274, 275]).

The first applications of interval methods concerned mainly the problem of assuring the correctness and reliability of computations in view of rounding errors resulting from imperfect machine representation of real numbers. Over many years, this direction of development of interval analysis was dominant. But keeping track of rounding and truncation errors is not the only reason for using interval arithmetic in solving various problems of numerical computing. In the early 1960s, interval arithmetic became increasingly widespread, which is attributed to the appearance of Moore's Ph.D. thesis [145] and his famous book on *interval analysis* [146]. These works have extended the utility of interval arithmetic to solving problems involving uncertainties resulting not only from rounding and truncation errors, but also from limited data, unknown physical relationships, measurement (observational) errors, and various other kinds of inexact knowledge, i.e., from the so-called *epistemic uncertainty* (also termed as type B uncertainty, ignorance, incertitude, reducible uncertainty, or subjective uncertainty), which—contrast with *aleatory uncertainty* —might be reduced with additional data/information, or better modeling and better parameter estimation. However, the role of interval analysis in modeling of uncertain and incomplete data is important also for its strong relation with other methods of representation of qualitative knowledge. Intervals are considered in the framework of order of magnitude reasoning where the entire real line is divided into intervals using the so-called *landmarks*. In one of the most popular approaches, the real line is divided into three intervals: negative numbers (interval $(-\infty, 0)$), zero (degenerate interval $[0, 0]$), and positive numbers (interval $(0, +\infty)$). The characteristic of this structure was described in the works of Parsons [170], Schwartz and Chen [238], and Struss [264], for example. Interval analysis is also one of the

[1]The idea of representing an imprecise real number by its bounds is in fact quite old. It can be traced back to the third century BC, when Archimedes (287–212BC) wrote his work *On the measurement of the circle* [11]. He used the concept of two-sided bounds for the irrational number π. The result, which he found by approximating the circle with the inscribed and circumscribed 96-side regular polygons, was that "the ratio of the circumference to the diameter $< 3\frac{1}{7}$ and $> 3\frac{10}{71}$", i.e., $\pi \in (3\frac{1}{7}, 3\frac{10}{71})$.

basic techniques used in solving problems of fuzzy logic. Most of the works on fuzzy sets and fuzzy techniques include a chapter devoted to interval analysis (see, e.g., Klir and Yuan [101], Tsoukalas and Uhrig [268] and the references therein). The usefulness of interval arithmetic is justified by the fact that fuzzy logic objects (fuzzy sets, fuzzy numbers) can be viewed as a nested family of intervals called $\alpha-$cuts of fuzzy sets (fuzzy numbers) (Buckley and Qu [21]). Many operations on fuzzy sets can be formulated using $\alpha-$cuts. Intervals can also be considered as a special example of a membership function. Very often, it is difficult to obtain an exact value of the latter. It seems more natural that an expert gives an interval of admissible values. Such an approach resulted in type-2 fuzzy sets that were successfully used by Kohout, Mendel, and Turksen (cf. Zadeh *et al.* [277]). It is worth mentioning that Turksen is one of the pioneers in the domain of using interval methods in expert systems and knowledge representation. Furthermore, intervals are a special case of *complex convex sets*. There is also a strong correlation between interval analysis and *convex analysis*. Interval analysis is also an important example of *set-valued analysis*.

The ability of interval analysis to produce guaranteed solutions in the presence of both rounding errors and data uncertainty has resulted in the fact that the role of interval computing has continuously increased in numerical analysis. Also, there has been increased interest in solving linear systems whose coefficients and right-hand side vary in some real intervals. The first methods for solving such systems (see, e.g., [9, 69, 76, 120, 165, 167, 166, 236, 159, 160, 207, 225, 226] and the references given therein) assumed that the coefficients of a system vary independently between their lower and upper bounds; i.e., they were concerned with finding enclosures for the set of solutions to all real systems included in the interval system. In practical applications, however, the coefficients of a system often functionally depend on a set of parameters that vary between their lower and upper bounds. So, when solving such systems, it is reasonable to take these dependencies into account, i.e., to consider the solutions of only those systems that have a specific structure. Obviously, the “truncated” solution set is included in the solution set of the corresponding interval linear system, but can be very small compared to the latter (see Sect. 4.3).

Systems whose entries are functions of parameters varying within given intervals are called *parametric interval linear systems* (PILS) (sometimes also known as *parametrized interval linear systems* or *linear parametric systems*); the solution set of such systems (the aforementioned “truncated” solution set) is called a *parametric solution set*. In view of the fact that the parametric solution set is usually much smaller than the solution of the corresponding nonparametric interval linear system, methods for computing rigorous enclosures for the parametric solution set are of high value. Generally, we can distinguish parametric interval linear systems with *affine–linear* and *nonlinear dependencies*. The simpler (but still quite complex) affine–linear case has been extensively studied over past years (see, e.g., [47, 80, 107, 113, 172, 173, 174, 187, 189, 229, 245, 247, 283]), starting from the work of Jansson [90] who considered interval linear systems with symmetric matrices, skew-symmetric matrices, and dependencies in the right-hand side. There are, however, only a few works on solving parametric interval linear systems with nonlinear dependencies (cf. [45, 60, 108, 178, 251, 253]). Hence, in view of all the above, there should be no doubts about the importance of developing new technical means for solving problems involving parametric interval vectors and matrices. In addition, interval linear algebra is a basis for solving more complex interval-valued problems, such as interval mathematical programming, interval least squares, or statistics on interval data.

Goals of the Book

The main goal of this book is to present a framework for solving a general class of linear systems with coefficients being continuous functions of parameters varying within prescribed intervals. In particular, several new algorithms delivering sharp rigorous bounds for the solutions of such systems have been developed and implemented. Another goal is to provide a comprehensive overview of the theory relating the problem of solving *parametric interval linear systems* and to investigate the basic properties of *parametric interval matrices*.

Contents of the Book

This book consists of seven chapters. Chapter 1 introduces preliminary theory of interval arithmetic. A large part of this chapter is devoted to the best known methods for bounding the range of functions over a box. The main reason for paying so much attention to this subject is that all enclosure methods that are based on interval arithmetic are eventually enclosure methods for the range of some function. The chapter is concluded with a short discussion on the rigorous computer implementation of interval arithmetic.

Chapter 2 starts from a brief introduction to affine arithmetic. Next, *revised affine forms*, which constitute the basis for the proposed algorithmic framework, are presented. The revised affine forms seem to be less recognized than classical affine forms; however, they are better suited for performing complex calculation in our opinion. The large part of this chapter is devoted to the multiplication of revised affine forms, since this constitutes the main bottleneck of affine computation. Various methods for multiplication of revised affine forms are therefore developed and compared in terms of accuracy and time complexity. In particular, an efficient algorithm for Chebyshev minimum-error multiplication is presented, which was recently developed by the author of this book in cooperation with Hladík [258]. The algorithm has asymptotic linear time complexity, which is optimal. The lower complexity of affine operations should open new possibilities for using revised affine forms in solving practical problems.

In order to make this book accessible to larger audience, the relevant background material on interval matrices opens Chap. 3. Next, parametric interval matrices with both affine–linear and nonlinear dependencies are defined. An affine transformation is then proposed which reduces a parametric interval matrix to the normalized form, i.e., to a matrix of revised affine forms. Despite some loss of information, the proposed transformation significantly simplifies the computation and enables parametric interval matrices with both affine–linear and nonlinear dependencies to be treated in a unified manner. Since most of the existing methods assume affine–linear dependencies, the proposed affine transformation extends the applicability of these methods. Further in this chapter, several verifiable sufficient conditions for checking basic properties of parametric interval matrices (such as regularity, strong regularity, and stability) are discussed.

Chapter 4 is devoted to the problem of solving parametric interval linear systems with arbitrary dependencies. Various types of parametric solution sets are first discussed. Next, some visualizations of two- and three-dimensional parametric sets are provided in order to show the complexity of the considered problems and to emphasize consequences of neglecting data dependency. It is also shown that the problem of solving real, complex, over-, and under-determined systems can be reduced to the problem of solving square parametric interval linear systems; thus, all these systems can be treated in a unified manner.

Chapter 5 gives an overview of methods for solving parametric interval linear systems. Three classes of methods are distinguished. Methods from the first class deliver outer interval enclosures for the united parametric solution set. The second class includes methods for computing the parametric (p-)solution given as a vector of revised affine forms. The p-solution enables us to obtain both outer and inner bounds for the united solution set. The methods from the third class are global optimization methods that aim to deliver exact interval solutions. Additionally, several generalizations and improvements of selected methods are suggested. A common feature of all presented algorithms is the combination of floating-point arithmetic with revised affine forms, which yields powerful algorithms with automatic result verification. All the methods presented in this chapter are compared in terms of speed and accuracy by using some benchmark problems and practical problems from structural and electrical engineering.

Finally, Chap. 6 is devoted to the special case of parametric linear programming problem. Despite some simplifications, the problem is still extremely complex. The main result in this chapter is a method for solving the considered parametric linear programming problem, which is not a priori exponential. The method was recently elaborated by Kolev and Skalna [118].

This book is intended not only for researchers interested in developing rigorous methods of numerical linear algebra, but also for engineers dealing with problems involving uncertain data. The presented theory might also be useful in various other fields of numerical analysis, in computer graphics, economics, computational geometry, computer-aided design, computer-assisted proofs, computer graphics, control theory, solving constraint satisfaction problems, and global optimization.

Ackowledgements

The author would like to express her sincere gratitude to Prof. Janusz Kacprzyk for his encouragement to publish the book. The author would also like to thank Prof. Jan Tadeusz Duda for his great support and valuable advices. Many thanks are directed to Prof. Milan Hladík for his constructive comments that helped me to improve this book. Thanks are also due to Prof. Kolev, Dr. Jerzy Duda, Dr. Andrzej Pownuk, Dr. M. V. Rama Rao for their valuable cooperation. The author asks for forgiveness for any possible omissions.

Chapter 1
Interval Arithmetic

Significant discrepancies (between the computed and the true result) are very rare, too rare to worry about all the time, yet not rare enough to ignore

(Kahan)

The first applications of interval arithmetic and interval methods mainly concerned the problem of assuring the correctness and reliability of numerical computations, in view of rounding errors resulting from the imperfect machine representation of real numbers. Because of this representation, two types of rounding errors can occur. The first one follows from the approximation of real input data by machine numbers, and the second is due to the approximation of intermediate results by machine numbers. As an example of the latter, consider quadratic equation $x^2 + 2px - 1 = 0$. The roots of a quadratic equation are usually computed by using the following quadratic formula: $x_{1,2} = -p \pm \sqrt{p^2 + 1}$. In IEEE 754 *floating-point (FP) arithmetic*, which is the most-widely-used form of computer arithmetic, if $1/p^2 \leqslant \mathbf{u}$, where $\mathbf{u}$ is *machine epsilon* (half[1] the distance between 1 and its successor), then $p^2 + 1 = p^2$, and we obtain $x_1 = -p + p = 0$ in computer arithmetic (which is incorrect). Since $\mathbf{u} = 2^{-53}$ in `binary64` floating-point representation, the `binary64` floating-point arithmetic will produce the same incorrect result for each $p \geqslant 10^8$. What is worse, no warning will be displayed, so an inexperienced user will trust the result. In order to manage the problem, the root of the smaller magnitude should be computed via equation $x_1 = q/\sqrt{p^2 + q}$ $(p > 0)$. Then, $x_1 = 5 \cdot 10^{-9}$ is produced, which is the correct value. Interval arithmetic in this case is not very helpful, since it yields range $[0, 1.4901 \cdot 10^{-8}]$ for x_1. This range takes rounding errors into account, of course,

[1] IEEE standard does not define the term *machine epsilon*, therefore, an alternative definition to the one given above can be found in various sources that identifies machine epsilon with the *unit in the last place*, i.e., the smallest representable positive normal floating-point number.

I. Skalna, *Parametric Interval Algebraic Systems*, Studies in Computational Intelligence 766, https://doi.org/10.1007/978-3-319-75187-0_1

but it is not very informative. In [234] it is argued that the wide range, such as the one obtained for x_1, brings no information about the sensitivity of the problem. In the general case, this is true since the wide range may be attributed to overestimation, which is inherent to interval computation. In this example, however, the wide range does indicate the sensitivity of the problem since the computation is relatively simple and the input data is crisp. Interval arithmetic "tells" us, in this case, that we should rearrange the computation or use higher-precision arithmetic. Several other examples of the pitfalls of floating-point processing can be found in [53, 148, 200, 234], for example. One of the best-known examples of the catastrophic consequences of rounding errors is the failure of the MIM-104 Patriot missile defense system during Operation Desert Storm at Dahram, Saudi Arabia, on Feb 25, 1991 [284]. Interval arithmetic, if carefully implemented, can be useful in tracking rounding errors in numerical computing. However, like any other tool, it is not a panacea for all problems in scientific computing. If used inappropriately, the results might be useless [234].

Over many years, the ability of interval arithmetic to cope with rounding errors was considered as the main reason for its "existence." This view changed in the 1960s with the appearance of Moore's book on interval analysis [146]. Some consider this book as the birth of modern interval arithmetic. According to Moore, besides bounding the effects of rounding and truncation errors in finite precision arithmetic, interval arithmetic can probe the behavior of functions efficiently and reliably over a whole set of arguments at once [4, 146]. Soon, it turned out that interval arithmetic is extremely useful as a language for describing a class of uncertainties such as *bounded disturbances*, the *bounded error approach*, and the *bounded parameter model*.

Starting in the mid-nineties, the interval approach has been used to describe parameter uncertainty in engineering systems. A typical example of the application of interval arithmetic in engineering practice is the study of the sensitivity of a system's behavior with respect to changes in input parameters (see, e.g., [153, 192, 246, 278]). The interval approach also turned out to be useful in economics [136], computer graphics [49, 152, 273], computer-assisted proofs [129], existence verification and construction of robust controllers [157], and control theory [19, 68, 68]. The range of applications of interval arithmetic is far broader (see, e.g., [29, 99]) and continues to grow.

The role of interval arithmetic in the modeling of uncertain and incomplete data is also important for its strong relationship with other methods of representation of uncertain knowledge. Interval analysis is one of the basic techniques used in solving problems of fuzzy logic. Moreover, there is a strong correlation between interval analysis and *convex analysis* or *set-valued analysis*. Further motivation for using interval arithmetic is its intimate connection with random set theory, the Dempster-Shafer structure, and the probability-bounds approach. Analysis of these uncertainty models and interval analysis are mutually relevant.

In the next section, we present the basic concepts of interval arithmetic. In Sect. 1.3, we discuss the best-known methods for computing the range of multivariate functions on box bounding, since range bounding is of great importance for the problem of solving parametric interval linear systems (PILS).

1.1 Intervals

A real (bare[2]) *interval* $\boldsymbol{x}$ is a closed and connected subset of real numbers (in a topological sense) defined by

$$\boldsymbol{x} \equiv [\underline{x}, \overline{x}] = \{ x \in \mathbb{R} \mid \underline{x} \leqslant x \leqslant \overline{x} \}, \tag{1.1}$$

where $\underline{x}$, $\overline{x} \in \overline{\mathbb{R}}$ denote, respectively, the *lower bound* and *upper bound* of $\boldsymbol{x}$. It is assumed that, for nonempty interval $\boldsymbol{x}$, its bounds $\underline{x}$ and $\overline{x}$ satisfy the following inequalities[3]: $\underline{x} \leqslant \overline{x}$, $\underline{x} < +\infty$ and $\overline{x} > -\infty$. If unknown quantity $\tilde{x}$ is represented by interval $\boldsymbol{x}$, then the only information available is that $\tilde{x} \in [\underline{x}, \overline{x}]$, without any further knowledge about which value from $\boldsymbol{x}$, is the most-likely value of $\tilde{x}$.

The set of all real intervals is denoted by $\overline{\mathbb{IR}}$. It is comprised of the empty interval, denoted by Ø or `Empty`, together with all nonempty real intervals. Since an interval is a set, set operations are well-defined on $\overline{\mathbb{IR}}$. Real numbers are embedded into $\overline{\mathbb{IR}}$ by the injection, which assigns interval $[x, x]$ of zero width to real number x.

The endpoints of an interval are often described by using the inf and sup operations, hence, the name of the *inf-sup* representation. The inf and sup operations are especially useful when, instead of $\boldsymbol{x}$, a more-complex expression occurs. According to the IEEE Std 1788–2015 [52]

$$\inf(\boldsymbol{x}) = \begin{cases} \underline{x} & \text{if } \boldsymbol{x} \text{ is nonempty,} \\ +\infty & \text{otherwise.} \end{cases} \tag{1.2}$$

$$\sup(\boldsymbol{x}) = \begin{cases} \overline{x} & \text{if } \boldsymbol{x} \text{ is nonempty,} \\ -\infty & \text{otherwise.} \end{cases} \tag{1.3}$$

Note: the definitions of numeric functions of an interval given below are also consistent with the IEEE Std 1788–2015 [52].

Motivated by the fact that unknown quantity $\tilde{x}$ can be represented as an approximation plus/minus a maximal error bound, the *midpoint*

$$x^c = \text{mid}(\boldsymbol{x}) = \begin{cases} (\underline{x} + \overline{x})/2 & \text{if } \boldsymbol{x} \text{ is nonempty bounded,} \\ \text{no value} & \text{if } \boldsymbol{x} \text{ is empty or unbounded.} \end{cases} \tag{1.4}$$

and the *radius*

$$x^{\Delta} = \text{rad}(\boldsymbol{x}) = \begin{cases} (\overline{x} - \underline{x})/2 & \text{if } \boldsymbol{x} \text{ is nonempty,} \\ \text{no value} & \text{if } \boldsymbol{x} \text{ is empty.} \end{cases} \tag{1.5}$$

[2]The term bare is used to emphasize that we mean the *non-decorated interval*. A *decorated interval* is an ordinary interval tagged with a few bits that encode the decoration [52].

[3]This definition, introduced by the IEEE 1788–2015 Standard for Interval Arithmetic [52], implies that $-\infty$ and $+\infty$ can be the bounds of an interval but can never belong to it. In particular, interval $[-\infty, +\infty]$ denotes the whole real line $\mathbb{R}$ (Entire), not the whole extended real line $\overline{\mathbb{R}}$.

of an interval $\boldsymbol{x}$ are introduced. Twice the radius gives the *width* of interval

$$\operatorname{wid}(\boldsymbol{x}) = \begin{cases} \overline{x} - \underline{x} & \text{if } \boldsymbol{x} \text{ is nonempty,} \\ \text{no value} & \text{if } \boldsymbol{x} \text{ is empty.} \end{cases} \tag{1.6}$$

Using the midpoint and radius, interval $\boldsymbol{x}$ can be written in the so-called *mid-rad* form[4]

$$\boldsymbol{x} = x^c + x^{\Delta}[-1, 1] \tag{1.7}$$

symbolically denoted as

$$\boldsymbol{x} \equiv \langle x^c, x^{\Delta} \rangle.$$

It is not hard to see that the following relationship holds:

$$x \in \boldsymbol{x} \Leftrightarrow |x - x^c| \leqslant x^{\Delta}.$$

The *absolute value* or *modulus* can be extended to intervals in several ways. The most-useful extension is the *magnitude* of $\boldsymbol{x}$ that is the maximal absolute value of the elements of $\boldsymbol{x}$:

$$|\boldsymbol{x}| = \operatorname{mag}(\boldsymbol{x}) = \begin{cases} \sup\{\ |x| \mid \underline{x} \leqslant x \leqslant \overline{x}\ \} & \text{if } \boldsymbol{x} \text{ is nonempty,} \\ \text{no value} & \text{if } \boldsymbol{x} \text{ is empty.} \end{cases} \tag{1.8}$$

Another relevant extension is the *mignitude* of $\boldsymbol{x}$, defined as:

$$\langle \boldsymbol{x} \rangle = \operatorname{mig}(\boldsymbol{x}) = \begin{cases} \inf\{\ |x| \mid \underline{x} \leqslant x \leqslant \overline{x}\ \} & \text{if } \boldsymbol{x} \text{ is nonempty,} \\ \text{no value} & \text{if } \boldsymbol{x} \text{ is empty.} \end{cases} \tag{1.9}$$

For nonempty interval $\boldsymbol{x}$, the magnitude and mignitude can be expressed using the following endpoint representation:

$$|\boldsymbol{x}| = \max\{\ |\underline{x}|, |\overline{x}|\ \},$$
$$\langle \boldsymbol{x} \rangle = \begin{cases} \min\{\ |\underline{x}|, |\overline{x}|\ \} & \text{if } 0 \notin \boldsymbol{x}, \\ 0 & \text{if } 0 \in \boldsymbol{x}. \end{cases}$$

Elementary arithmetic operations $\bullet \in \Omega = \{+, -, *, /\ \}$ are defined on the set of intervals by

$$\boldsymbol{x} \bullet \boldsymbol{y} = \operatorname{hull}\left(\{\ x \bullet y \mid x \in \boldsymbol{x}, y \in \boldsymbol{y}, x \bullet y \text{ is defined }\}\right), \tag{1.10}$$

where $\operatorname{hull}(S)$, $S \subseteq \mathbb{R}^n$, is the tightest member of $\overline{\mathbb{IR}}^n$ that contains S, i.e.,

$$\operatorname{hull}(S) = \bigcap \left\{ Y \in \overline{\mathbb{IR}}^n \mid S \subseteq Y \right\} = [\inf S, \sup S]. \tag{1.11}$$

[4]The mid-rad representation is not included in the IEEE Std P1788-2015.

The intersection of two non-empty intervals $\boldsymbol{x}$, $\boldsymbol{y}$ is defined as

$$\boldsymbol{x} \cap \boldsymbol{y} = \begin{cases} [\max\{\underline{x}, \underline{y}\}, \min\{\overline{x}, \overline{y}\}] & \text{if non-empty}, \\ \text{no value} & \text{otherwise}. \end{cases} \tag{1.12}$$

Equation (1.10) restricts the division to intervals $\boldsymbol{y}$ with $0 \notin \boldsymbol{y}$.[5] Moreover, it immediately follows from (1.10) that the image of each of the four operations is the exact image of the respective real operations and that

$$\tilde{x} \in \boldsymbol{x},\ \tilde{y} \in \boldsymbol{y} \Rightarrow \tilde{x} \bullet \tilde{y} \in \boldsymbol{x} \bullet \boldsymbol{y}, \tag{1.13}$$

which means that the unknown exact result of an operation is enclosed by the resulting interval. Property (1.13) is called the *inclusion principle of interval arithmetic*.

Another important property of interval arithmetic (often considered fundamental) is that, for arbitrary intervals $\boldsymbol{x}$, $\boldsymbol{y}$, $\boldsymbol{u}$, $\boldsymbol{v}$, it holds that

$$\boldsymbol{x} \subseteq \boldsymbol{y} \wedge \boldsymbol{u} \subseteq \boldsymbol{v} \Rightarrow \boldsymbol{x} \bullet \boldsymbol{u} \subseteq \boldsymbol{y} \bullet \boldsymbol{v}, \tag{1.14}$$

if the operation is defined. Property (1.14) is called the *inclusion isotonicity* or *inclusion monotonicity*.

Equation (1.10) is not very convenient in practical computations. Moore [145] proved that arithmetic operations can be defined using the endpoints of their operands. Given two nonempty intervals $\boldsymbol{x}$ and $\boldsymbol{y}$, we have

$$\begin{aligned} \boldsymbol{x} + \boldsymbol{y} &= [\underline{x} + \underline{y},\ \overline{x} + \overline{y}], \\ \boldsymbol{x} - \boldsymbol{y} &= [\underline{x} - \overline{y},\ \overline{x} - \underline{y}], \\ \boldsymbol{x} * \boldsymbol{y} &= [\min\{\underline{x}\underline{y},\ \underline{x}\overline{y},\ \overline{x}\underline{y},\ \overline{x}\overline{y}\},\ \max\{\underline{x}\underline{y},\ \underline{x}\overline{y},\ \overline{x}\underline{y},\ \overline{x}\overline{y}\}], \\ \boldsymbol{x}/\boldsymbol{y} &= [\underline{x},\ \overline{x}] \cdot [1/\overline{y},\ 1/\underline{y}],\ 0 \notin \boldsymbol{y}. \end{aligned} \tag{1.15}$$

More-advanced operations on intervals such as the square, square root, exponential, logarithm, sinus, cosinus, and arcus tangens can be expressed for intervals arguments in terms of endpoints and the local minima and maxima.

The distance between intervals $\boldsymbol{x}$ and $\boldsymbol{y}$ is defined by the *Hausdorff distance* [147]:

$$d_H(\boldsymbol{x}, \boldsymbol{y}) = \max\{|\underline{x} - \underline{y}|, |\overline{x} - \overline{y}|\}. \tag{1.16}$$

It can be easily verified that d_H is a metric, so $(\mathbb{IR}, d_H)$ is a metric space [160]. In the metric topology induced by d_H, the midpoint, radius, magnitude, mignitude, infimum, supremum, and elementary arithmetic operations are continuous functions of their arguments in the usual sense (as long as no division by an interval containing zero occurs) [147].

---

[5] Some modifications of the interval division can be found in the literature [93, 199]. They result from the fact that the restriction imposed on the divisor is acknowledged to be unacceptable and unnecessary by several authors.

Order relations $\omega \in \{<, \leqslant, >, \geqslant\}$ are extended to interval quantities by defining [160]:

$$\boldsymbol{x}\, \omega\, \boldsymbol{y} \Leftrightarrow \tilde{x}\, \omega\, \tilde{y}, \text{ for all } \tilde{x} \in \boldsymbol{x}, \tilde{y} \in \boldsymbol{y}.$$

It is not hard to see that

$$\boldsymbol{x} < \boldsymbol{y} \Leftrightarrow \overline{x} < \underline{y},\ \boldsymbol{x} > \boldsymbol{y} \Leftrightarrow \underline{x} > \overline{y},$$
$$\boldsymbol{x} \leqslant \boldsymbol{y} \Leftrightarrow \overline{x} \leqslant \underline{y},\ \boldsymbol{x} \geqslant \boldsymbol{y} \Leftrightarrow \underline{x} \geqslant \overline{y}.$$

The order relations are antisymmetric and transitive, additionally $\leqslant$ and $\geqslant$ are reflexive. However, two intervals might not be comparable, e.g., $[1, 3] \nleqslant [2, 4] \nleqslant [1, 3]$.

1.2 Basic Properties of Interval Arithmetic

The algebraic system of interval arithmetic is only an *abelian monoid* under both addition and multiplication, additive and multiplicative inverses exist only for intervals of zero radius. For example, suppose that $\boldsymbol{x}$ is such that $x^{\Delta} > 0$ and $\boldsymbol{a}$ is its additive inverse. Then,

$$\begin{aligned}
&[\underline{x}, \overline{x}] - [\underline{a}, \overline{a}] = 0 \ \wedge\ \underline{x} < \overline{x} \ \wedge\ \underline{a} \leqslant \overline{a} \Rightarrow \\
&[\underline{x} - \overline{a}, \overline{x} - \underline{a}] = 0 \ \wedge\ \underline{x} < \overline{x} \ \wedge\ \underline{a} \leqslant \overline{a} \Rightarrow \\
&\underline{x} = \overline{a} \ \wedge\ \overline{x} = \underline{a} \ \wedge\ \underline{x} < \overline{x} \ \wedge\ \underline{a} \leqslant \overline{a} \Rightarrow \\
&\underline{a} < \overline{a} \ \wedge\ \overline{a} \leqslant \underline{a} \text{ which is a contradiciton.}
\end{aligned}$$

Now, let $0 \notin \boldsymbol{x}$, $x^{\Delta} > 0$, and let $\boldsymbol{a}$ be the inverse of $\boldsymbol{x}$. Without loss of generality we can assume that $\underline{x} > 0$. Then, $\underline{a} > 0$ and

$$\begin{aligned}
&[\underline{x}, \overline{x}] * [\underline{a}, \overline{a}] = 1 \ \wedge\ \underline{x} < \overline{x} \ \wedge\ \underline{a} \leqslant \overline{a} \Rightarrow \\
&\overline{a} = 1/\overline{x} < 1/\underline{x} = \underline{a} \ \wedge\ \underline{a} \leqslant \overline{a} \text{ which is a contradiciton.}
\end{aligned}$$

Apart from the lack of inverse elements, interval arithmetic operations only preserve some of the algebraic laws valid for real numbers. In particular, the following laws hold true for arbitrary intervals $\boldsymbol{x}$, $\boldsymbol{y}$, and $\boldsymbol{z}$:

$$\begin{array}{lll}
\boldsymbol{x} + \boldsymbol{y} = \boldsymbol{y} + \boldsymbol{x}, & \boldsymbol{x}\boldsymbol{y} = \boldsymbol{y}\boldsymbol{x}, & \text{(commutativity)} \\
(\boldsymbol{x} + \boldsymbol{y}) + \boldsymbol{z} = \boldsymbol{x} + (\boldsymbol{y} + \boldsymbol{z}), & (\boldsymbol{x}\boldsymbol{y})\boldsymbol{z} = \boldsymbol{x}(\boldsymbol{y}\boldsymbol{z}), & \text{(associativity)} \\
\boldsymbol{x} + 0 = 0 + \boldsymbol{x}, & \boldsymbol{x} * 1 = 1 * \boldsymbol{x}, & \text{(neutral element)} \\
\boldsymbol{x} - \boldsymbol{y} = \boldsymbol{x} + (-\boldsymbol{y}) = -\boldsymbol{y} + \boldsymbol{x}, & \boldsymbol{x}/\boldsymbol{y} = \boldsymbol{x} * \boldsymbol{y}^{-1} = \boldsymbol{y}^{-1} * \boldsymbol{x}, & \\
-(\boldsymbol{x} - \boldsymbol{y}) = \boldsymbol{y} - \boldsymbol{x}, & \boldsymbol{x}(-\boldsymbol{y}) = (-\boldsymbol{x})\boldsymbol{y} = -\boldsymbol{x}\boldsymbol{y}. &
\end{array}$$

The distributivity and cancellation are present in the interval arithmetic in a weaker form as *sub-distributivity* and *sub-cancellation*, respectively [160]. Thus, for arbitrary interval $\boldsymbol{x}$, $\boldsymbol{y}$, $\boldsymbol{z}$, the following holds true:

$$\begin{array}{lll} \boldsymbol{x}(\boldsymbol{y} \pm \boldsymbol{z}) \subseteq \boldsymbol{x}\boldsymbol{y} \pm \boldsymbol{x}\boldsymbol{z}, & (\boldsymbol{x} \pm \boldsymbol{y})\boldsymbol{z} \subseteq \boldsymbol{x}\boldsymbol{z} \pm \boldsymbol{y}\boldsymbol{z}, & \text{(sub-distributivity)} \\ \left.\begin{array}{ll} \boldsymbol{x} - \boldsymbol{y} \subseteq (\boldsymbol{x} + \boldsymbol{z}) - (\boldsymbol{y} + \boldsymbol{z}), & \boldsymbol{x}/\boldsymbol{y} \subseteq (\boldsymbol{x}\boldsymbol{z})/(\boldsymbol{y}\boldsymbol{z}), \\ 0 \in \boldsymbol{x} - \boldsymbol{x}, & 1 \in \boldsymbol{x}/\boldsymbol{x}. \end{array}\right\} & & \text{(sub-cancellation)} \end{array}$$

The lack of distributivity and cancellation is due to the fact that interval arithmetic does not identify different occurrences of the same variable in an expression. In the next section, we will discuss this problem in the context of range computation.

1.3 Range Bounding

The problem of range bounding is often called the *main problem of interval computation* [91, 99, 146]. Bounding the range of real-valued multivariate (continuous) functions over a box is of great importance for the problem of solving parametric interval linear systems, since all interval methods for solving such systems are eventually enclosure methods for the range of real-valued functions. Range bounding is also very important for the problem of localizing and enclosing global minima (cf. [73]), fixed point theorems (cf. [229]), verifying and enclosing solutions of initial value problems, and verifying the (non)existence of zeros of a function in a box (cf. [5, 120, 121, 146]).

The problem of computing the range can be formalized as follows: given function $f : D \subseteq \mathbb{R}^n \to \mathbb{R}$ and box $\boldsymbol{x}$, we are interested in range

$$\mathrm{Rge}(f \mid \boldsymbol{x}) = \{f(x) \mid x \in \boldsymbol{x} \cap D\}. \tag{1.17}$$

If f is continuous, then Rge $(f \mid \boldsymbol{x})$ is a closed interval. The *set-valued* mapping that assigns range Rge $(f \mid \boldsymbol{x})$ to each $\boldsymbol{x} \subset \mathbb{R}^n$ is called the *united extension* of f [147].

However, for many classes of functions, computing interval range Rge $(f \mid \boldsymbol{x})$ is an NP-hard problem (see, e.g., [54, 97, 99, 124, 147, 270]). What is more, it remains NP-hard even if we want to compute the range within a given accuracy $\varepsilon > 0$ [125]. Therefore, in practical situations, we often ask for an *interval enclosure* for Rge $(f \mid \boldsymbol{x})$, i.e., an interval $\boldsymbol{y}$ such that Rge $(f \mid \boldsymbol{x}) \subseteq \boldsymbol{y}$.

If real function f is represented by arithmetic expression f, then $\boldsymbol{y}$ can be obtained by a (straightforward) *interval evaluation* [160] of f over box $\boldsymbol{x}$. The interval evaluation relies on replacing the real variables by the corresponding interval variables and the real arithmetic operations by the corresponding interval operations. Interval function $\boldsymbol{f}$, induced by expression f, is said to be the *natural interval extension* of f.

If the interval evaluation is defined, then $f(x) \in \boldsymbol{f}(\boldsymbol{x})$. The interval evaluation is not defined if $\boldsymbol{x}$ contains a point $\tilde{x} \notin D_f$. Sometimes, the interval evaluation is not defined even if $\boldsymbol{x} \subseteq D_f$. The result depends on the syntactic formulation of the

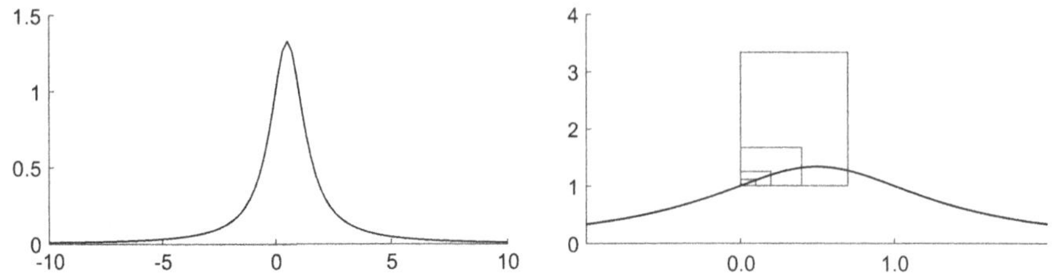

Fig. 1.1 Graph of function f described by expression $\mathsf{f}(x) = 1/\left(1 - x + x^2\right)$ (left); enclosures for Rge $(f \mid [0, t])$, $t = 0.1, 0.2, 0.3, 0.4$ (right)

expression for f. If, for example, f is defined by expression $\mathsf{f}_1(x) = \frac{1}{(x \cdot x + 2)}$, then $\boldsymbol{f}([-2, 2])$ is not defined because the denominator contains 0 ($[-2, 2] \cdot [-2, 2] + 2 = [-2, 6]$) whereas, if f is defined by $\mathsf{f}_2(x) = \frac{1}{(x^2+2)}$, then $\boldsymbol{f}([-2, 2]) = [1/6, 1/2]$.

Theorem 1.1 (cf. Neumaier [160]) *If interval function $\boldsymbol{f}$ is a natural interval extension of real function f, then $\boldsymbol{f}$ is inclusion monotonic.*

The natural interval extension can be obtained for any *computable function*.[6] The question is whether the obtained bounds are satisfactory. There is no simple answer to this question. A natural interval extension can yield the exact range, but it can also produce an arbitrarily large overestimation as is shown in the following example.

Example 1.2 (*[160]*) Consider real-valued function f (Fig. 1.1) described by arithmetic expression $\mathsf{f}(x) = 1/\left(1 - x + x^2\right)$. The function is defined for all real numbers and takes values from interval $(0, 4/3]$. However, for intervals $[0, t]$ where $0 \leqslant t < 1$, the natural interval extension of f yields range

$$\mathsf{f}([0, t]) = 1/\left(1 - [0, t] + [0, t]^2\right) = \left[1/\left(1 + t^2\right), 1/(1 - t)\right],$$

whose upper bound tends to $+\infty$ as t tends to 1^- (Fig. 1.1. This huge overestimation is caused by multiple occurrences of variable x in the considered expression. Subexpressions $1 - x$ and x^2 are calculated as if they were independent, i.e., x is treated as a different variable in each occurrence. As a consequence, the range of function f is calculated from expression $\mathsf{f}(x_1, x_2) = 1/\left(1 - x_1 + x_2^2\right)$ over box $[0, t] \times [0, t]$.

Multiple occurrences of one or several variables in an expression lead to the so-called *dependency problem* (*dependence problem, interval dependency*), which is inherent to interval computation (and which is the main source of overestimation in interval computations). The dependency problem is our main interest, since it is strictly connected with the problem of solving parametric interval linear systems.

In some cases, it is possible to transform an expression to an equivalent form with a single occurrence of each variable. In Example 1.2, we could replace expression f

[6] According to the Church–Turing thesis, any function that has an algorithm is computable. An algorithm in this sense is understood to be a sequence of steps a person with unlimited time and an infinite supply of pen and paper could follow.

by expression $\mathbf{g}(x) = 4/(3 + (1 - 2x)^2)$, which is equivalent to f in real numbers.[7] Interval evaluation of $\mathbf{g}$ for interval [0, 1] gives range

$$\mathbf{g}([0, 1]) = 4/(3 + (1 - 2[0, 1])^2) = 4/[3, 4] = [1, 4/3],$$

which is the exact range.

Theorem 1.3 (cf. Moore [146], Neumaier [160]) *Let $\boldsymbol{f}$ be a natural interval extension of real function f represented by arithmetic expression* f *in which each variable x_i $(i = 1, \ldots, n)$ occurs only once. Then, for all $\boldsymbol{x} \subseteq \mathrm{Dom}(f)$,*

$$\boldsymbol{f}(\boldsymbol{x}) = \mathrm{Rge}\,(f \mid \boldsymbol{x})\,.$$

Unfortunately, for many expressions, it is difficult to rearrange them so that the assumptions of Theorem 1.3 are fulfilled. If it is not possible to obtain a single occurrence of each variable, we should at least try to reduce the number of occurrences of all variables.

Theorem 1.4 (cf. Moore [146], Neumaier [160]) *Let $\boldsymbol{f}$ be a natural interval extension of real function f represented by arithmetic expression* f *in $n + m$ variables $(\xi_1, \ldots, \xi_n, \eta_1, \ldots, \eta_m) = (\xi, \eta)$, and assume that variables η_i $(i = 1 \ldots, m)$ occur only once in* f*. Then,*

$$\mathrm{Rge}\,(f \mid (\boldsymbol{x}, \boldsymbol{y})) = \bigcup_{\xi \in \boldsymbol{x}} \boldsymbol{f}(\xi, \boldsymbol{y}), \tag{1.18}$$

for all $\boldsymbol{x}$, $\boldsymbol{y}$, such that $(\boldsymbol{x}, \boldsymbol{y}) \subseteq \mathrm{Dom}\, f$.

Symbolic manipulation of arithmetic expressions [23] has proven to be useful in the detection and removal of multiple occurrences of variables. Obviously, we can never be sure to obtain the best form of an expression, but even a small reduction in the number of occurrences of variables will improve the accuracy of the final enclosure. For multivariate polynomials, Moore [149] proposed a number of rearrangement techniques.

However, the natural interval extension is only one of the infinitely many inclusion monotonic interval extensions of a real function. So, despite the reduction of the number of occurrences of variables, if the obtained interval enclosure is still not satisfactory, some more-sophisticated interval extension must be employed to reduce the effect of the interval dependency. Interval function $\boldsymbol{f}$ is said to be *interval extension* of real function f if (cf. [52, 160])

$$f(x) = \boldsymbol{f}(x), \forall x \in \mathrm{Dom}\,(f), \tag{1.19a}$$

$$f(x) \in \boldsymbol{f}(\boldsymbol{x}), \forall x \in \boldsymbol{x} \subseteq \mathrm{Dom}\,(f). \tag{1.19b}$$

[7] Floating-point arithmetic is not associative, therefore, two expressions equivalent for real arguments are not necessarily equivalent in floating-point arithmetic.

If $\boldsymbol{f}$ satisfies condition (1.19b) only, then it is called an *interval inclusion* function of f [122].

Theorem 1.5 (Fundamental Theorem of Interval Arithmetic (FTIA) [147]) *If interval function $\boldsymbol{f}$ is an inclusion monotonic interval extension of real function f, then, for each $\boldsymbol{x} \subseteq \mathrm{Dom}\,(f)$,*

$$\mathrm{Rge}\,(f \mid \boldsymbol{x}) \subseteq \boldsymbol{f}(\boldsymbol{x}). \tag{1.20}$$

The quality of the interval bounds is usually quantified using the *excess width* [147] $\mathrm{wid}(\boldsymbol{f}(\boldsymbol{x})) - \mathrm{wid}(\mathrm{Rge}\,(f \mid \boldsymbol{x}))$ and the order of convergence. We say that the interval extension has the kth order approximation property in $\boldsymbol{x}_0$ if, for each $\boldsymbol{x} \subseteq \boldsymbol{x}_0$, $\boldsymbol{f}$ is defined in $\boldsymbol{x}$ and

$$\mathrm{wid}(\boldsymbol{f}(\boldsymbol{x})) - \mathrm{wid}(\mathrm{Rge}\,(f \mid \boldsymbol{x})) \leqslant \kappa \|\mathrm{wid}(\boldsymbol{x})\|_{\infty}^{k}, \tag{1.21}$$

where κ is a real non-negative constant. A range enclosure method is said to have approximation order k if it has this order for (at least) all polynomials and all boxes (cf. [91, 97, 160, 161]).

Interval extension $\boldsymbol{f}$ of real function f is said to be *Lipschitz* in $\boldsymbol{x}_0 \subseteq \mathrm{Dom}(f)$ if there is a real constant L such that, for each $\boldsymbol{x} \subseteq \boldsymbol{x}_0$,

$$\mathrm{wid}(\boldsymbol{f}(\boldsymbol{x})) \leqslant L \|\mathrm{wid}(\boldsymbol{x})\|_{\infty}, \tag{1.22}$$

where constant L depends on $\boldsymbol{x}_0$ but not on $\boldsymbol{x}$.

Theorem 1.6 *Lipschitz interval extension $\boldsymbol{f}$ of real-valued function f converges linearly to the exact range of f.*

Proof If $\boldsymbol{f}$ is Lipschitz, then (1.22) holds for some constant L, and we have

$$\mathrm{wid}(\boldsymbol{f}(\boldsymbol{x})) - \mathrm{wid}(\mathrm{Rge}\,(f \mid \boldsymbol{x})) \leqslant \mathrm{wid}(\boldsymbol{f}(\boldsymbol{x})) \leqslant L \|\mathrm{wid}(\boldsymbol{x})\|_{\infty},$$

which gives the linear convergence. □

Probably the most-studied interval extensions of real functions (see, e.g., [15, 70, 71, 119, 147, 193–196, 198, 223, 224]) are *centered forms*, originally introduced by Moore [146] for polynomial and rational functions. In order to obtain a centered form, function f is written as (cf. [150, 193])

$$f(x) = f(c) + g(y), \tag{1.23}$$

where $y_i = x_i - c_i$ $(i = 1, \ldots, n)$, and g is defined by (1.23):

$$g(y) = f(y + c) - f(c). \tag{1.24}$$

It is assumed that the expression for g is as simple as possible. Corresponding centered form $\boldsymbol{f}_C$ is given by formula [193]

$$\boldsymbol{f}_C(\boldsymbol{x}) = f(c) + \boldsymbol{g}(\boldsymbol{x} - c), \tag{1.25}$$

where $c \in \boldsymbol{x}$ is a center point and $\boldsymbol{g}$ is an interval extension of g. Usually, c is chosen as the midpoint of $\boldsymbol{x}$, i.e., $c = x^c$.

A large number of centered forms are possible. The simplest centered form is the *mean value form*

$$\boldsymbol{f}_M(\boldsymbol{x}) = f(x^c) + \nabla \boldsymbol{f}(\boldsymbol{x})^T (\boldsymbol{x} - x^c), \tag{1.26}$$

where the ith element of gradient vector $\nabla \boldsymbol{f}$ is an inclusion monotonic interval extension of the ith partial derivative of f over $\boldsymbol{x}$.

Theorem 1.7 *The mean value form (1.26) is an interval extension of f.*

Proof Let x be an arbitrary box such that $\boldsymbol{x} \subseteq \text{Dom } f$. For all $x \in \boldsymbol{x}$, it holds that

$$f(x) = f(x) + \nabla f^T(x)(x - x) = \boldsymbol{f}_M([x, x]) = \boldsymbol{f}_M(x).$$

Now, by the mean value theorem, ξ exists for each $x \in \boldsymbol{x}$, with ξ_i $(i = 1, \ldots, n)$ lying in the line segment between x_i and x_i^c, such that

$$f(x) = f(x^c) + \nabla f^T(\xi)(x - x^c).$$

Since the elements of $\nabla \boldsymbol{f}$ are interval extensions of $\partial f / \partial x_i$ and interval arithmetic operations are inclusion monotonic,

$$f(x^c) + \nabla f^T(\xi)(x - x^c) \in f(x^c) + \nabla \boldsymbol{f}^T(\boldsymbol{x})(\boldsymbol{x} - x^c),$$

and we have $f(x) \in \boldsymbol{f}_M(\boldsymbol{x})$. □

Theorem 1.8 *Mean value form (1.26) is inclusion monotonic, i.e.,*

$$\boldsymbol{x} \subseteq \boldsymbol{y} \Rightarrow \boldsymbol{f}_M(\boldsymbol{x}) \subseteq \boldsymbol{f}_M(\boldsymbol{y}). \tag{1.27}$$

Proof Let $\boldsymbol{x} \subseteq \boldsymbol{y}$. From (1.26) and the mean value theorem, we have

$$\boldsymbol{f}_M(\boldsymbol{x}) = f(y^c) + \nabla f(\xi)^T (x^c - y^c) + \nabla \boldsymbol{f}(\boldsymbol{x})^T (\boldsymbol{x} - x^c),$$

where $\xi \in \boldsymbol{y}$. Since the elements of $\nabla \boldsymbol{f}$ are inclusion monotonic interval extensions of the partial derivatives $\partial f / \partial x_i$,

$$|\boldsymbol{f}_M(\boldsymbol{x}) - f(y^c)| \leqslant |\nabla \boldsymbol{f}(\boldsymbol{y})^T| \left(|x^c - y^c| + x^\Delta\right) \leqslant |\nabla \boldsymbol{f}(\boldsymbol{y})^T| y^\Delta$$

by ([160], (1.6.15)). Since $(f(y^c))^\Delta = 0$ and $(\boldsymbol{y} - y^c)^c = 0$, we have

$$(\boldsymbol{f}_M(\boldsymbol{y}))^c = f(y^c),$$
$$(\boldsymbol{f}_M(\boldsymbol{y}))^\Delta = |\nabla \boldsymbol{f}(\boldsymbol{y})^T| y^\Delta.$$

Hence,

$$|\boldsymbol{f}_M(\boldsymbol{x}) - (\boldsymbol{f}_M(\boldsymbol{y}))^c| \leqslant (\boldsymbol{f}_M(\boldsymbol{y}))^\Delta$$

and ([160], (1.6.15)) gives (1.27). □

Notice that the inclusion monotonicity of the mean value form does not hold if we replace midpoint x^c by an arbitrary point in $\boldsymbol{x}$ [15, 160].

Under some assumption on the range of the gradient of f [10], mean value form (1.26) has quadratic approximation property [25], which means that for narrow intervals it generally gives much-better enclosures than the natural interval extension, which only has a linear approximation property [146, 160].

A further improvement of the interval bounds can be obtained if we replace (1.26) by centered form [123]

$$\boldsymbol{f}(\boldsymbol{x}) = f(z) + s_f[\boldsymbol{x}, z](\boldsymbol{x} - z), \tag{1.28}$$

where $z \in \boldsymbol{x}$ and $s_f[\boldsymbol{x}, z]$ is an *interval slope*, i.e. an interval extension of slope function $s_f[x, z]$. There are different definitions of the slope function, the one below is due to Krawczyk [122].

Definition 1.9 Let two functions $f : D \subseteq \mathbb{R}^n \to \mathbb{R}^m$ and $h : D \times D \to \mathbb{R}^{m \times n}$ be given. If f satisfies the condition

$$f(x) - f(z) = s_f[x, z](x - z), \text{ for all } x, z \in D, \tag{1.29}$$

then $m \times n$ matrix $s_f[x, z]$ is a slope of f.

Under natural conditions, centered form (1.28) has a quadratic approximation property (see, e.g., Hansen [71]). If f is continuously differentiable, then (cf. [128, 160])

$$s_f[\boldsymbol{x}, z] \subseteq s_f[\boldsymbol{x}, \boldsymbol{x}] = \nabla \boldsymbol{f}(\boldsymbol{x}), \tag{1.30}$$

which means that centered forms with interval slopes usually yield smaller intervals than mean value forms.

Another possibility to improve the bounds is to use a *generalized centered form* defined by

$$\boldsymbol{f}_G(\boldsymbol{x}, c) = f(c) + \boldsymbol{L}(\boldsymbol{x}, c)(\boldsymbol{x} - c), \tag{1.31}$$

where $c \in \boldsymbol{x}$ is a center point and $\boldsymbol{L}$ is an interval extension of function $L : \mathbb{R}^n \times \mathbb{R}^n \to \mathbb{R}^n$ that is a Lipschitz function for f over $\boldsymbol{x} \subseteq \mathrm{Dom}\,(f)$, i.e., L fulfills condition (cf. [15])

$$f(x) - f(c) \in \boldsymbol{L}(\boldsymbol{x}, c)(x - c). \tag{1.32}$$

Under certain requirements for function L, generalized centered form (1.31) has a quadratic approximation property. If $c = x^c$, then generalized centered form (1.31) is inclusion monotonic, which is very important, e.g., in global optimization, since inclusion monotonicity implies that the approximations to the global extrema of function f over $\boldsymbol{x}$ improve when $\mathrm{wid}(\boldsymbol{x})$ decreases. However, as shown in [15, 160], the midpoint does not necessarily yield the best bounds. A center that yields the greatest lower bound (the lowest upper bound) of a centered form among all centers within a given interval is called an *optimal center* with respect to the lower bound (upper bound). A centered form evaluated at an optimal center is called an *optimal centered form* [15].

Assuming that $\boldsymbol{L}$ does not depend on c, Baumann [15] derived simple expressions for computing the optimal centers.

Theorem 1.10 (Baumann [15]) *Suppose that, for real function* $f : D \subseteq \mathbb{R}^n \to \mathbb{R}$ *there is a Lipschitz function* L *defined on* $\boldsymbol{x} \subseteq D$*, and let* $\boldsymbol{L}(\boldsymbol{x}) = [\underline{l}, \overline{l}]$. *Then, vectors* c^- *and* c^+ *defined by their components*

$$c_i^- = \begin{cases} \overline{x}_i & \text{if } \overline{l}_i \leqslant 0, \\ \underline{x}_i & \text{if } \underline{l}_i \geqslant 0, \\ (\overline{l}_i \underline{x}_i - \underline{l}_i \overline{x}_i)/\mathrm{wid}(\boldsymbol{l}_i) & \text{otherwise}, \end{cases} \tag{1.33a}$$

$$c_i^+ = \begin{cases} \underline{x}_i & \text{if } \overline{l}_i \leqslant 0, \\ \overline{x}_i & \text{if } \underline{l}_i \geqslant 0, \\ -(\underline{l}_i \underline{x}_i - \overline{l}_i \overline{x}_i)/\mathrm{wid}(\boldsymbol{l}_i) & \text{otherwise} \end{cases} \tag{1.33b}$$

belong to interval $\boldsymbol{x}$*, and for all* $c \in \boldsymbol{x}$*, centered form (1.31) satisfies*

$$\inf(\boldsymbol{f}_G(\boldsymbol{x}, c)) \leqslant \inf(\boldsymbol{f}_G(\boldsymbol{x}, c^-)), \tag{1.34a}$$

$$\sup(\boldsymbol{f}_G(\boldsymbol{x}, c)) \geqslant \sup(\boldsymbol{f}_G(\boldsymbol{x}, c^+)). \tag{1.34b}$$

By taking the intersection of optimal centered forms $\boldsymbol{f}_G(\boldsymbol{x}, c^-)$ and $\boldsymbol{f}_G(\boldsymbol{x}, c^+)$, we obtain enclosure

$$\boldsymbol{f}_B(\boldsymbol{x}) = \boldsymbol{f}_G(\boldsymbol{x}, c^-) \cap \boldsymbol{f}_G(\boldsymbol{x}, c^+), \tag{1.35}$$

which is called a *bicentered form* (see, e.g., [50, 160, 162]). Obviously, the bicentered forms have second order approximation property. In an important special case when $\boldsymbol{L} = \nabla \boldsymbol{f}$ and $0 \notin \nabla \boldsymbol{f}(\boldsymbol{x})$, so that f is monotone in each variable, the extremal values of f are attained at the vertices of $\boldsymbol{x}$, and we have equalities in (1.34a) and (1.34b). This means that $\boldsymbol{f}_B(\boldsymbol{x}) = \mathrm{Rge}\,(f \mid \boldsymbol{x})$.

Kearfott and Du [98] observed that the so-called *cluster effect* (which occurs in branch and bound methods for minimizing a function in a box, for example) is an inevitable consequence of the first- and second-order enclosure methods. As explained in [98] (see also [161]), the cluster effect occurs when many small boxes containing no solutions remain uneliminated if they are near the minimum. Processing these boxes often dominate the total effort of computing the minimum. Kearfott and Du [98] showed that enclosure methods of at least third-order are needed to eliminate the cluster effect. However, for ill-conditioned problems, even third-order methods can fail to eliminate the cluster effect [97, 161].

Enclosure methods of higher than second-order were first considered by Cornelius and Lohner for the one-dimensional case [31] (see also Alefeld and Lehmer [8], Alefed [3]). They proposed a method that, for real function f, allows us to compute an interval enclosure with theoretically arbitrary order of convergence $n > 0$. In practice, however, the convergence of order $n \leqslant 4$ can be achieved with little or moderate effort and can handle order $n = 5, 6$ with an increased effort [31]. The result of Cornelius and Lohner is based upon the assumption that a function can be written as

$$f(x) = f(c) + (x - c)^n h(x), \tag{1.36}$$

where $c \in \boldsymbol{x}$ and h is a real function continuous on $\boldsymbol{x}$. Then, the following centered form can be defined:

$$\boldsymbol{f}(\boldsymbol{x}) = f(c) + \text{Rge}\left((x - c)^n \mid \boldsymbol{x}\right) \boldsymbol{h}(\boldsymbol{x}), \tag{1.37}$$

where $\boldsymbol{h}$ is an interval extension of h.

Theorem 1.11 (cf. Alefeld [3], Alefeld and Lohner [8]) *Let $f : D \subseteq \mathbb{R} \to \mathbb{R}$ be a real function, $\boldsymbol{x} \subseteq D$, and $\boldsymbol{h}$ be an interval extension of a real continuous function h defined on $\boldsymbol{x}$. Then, centered form (1.37) is an interval extension of function f. If, for some $\sigma > 0$, $\boldsymbol{h}$ fulfills*

$$\text{wid}(\boldsymbol{h}(\boldsymbol{x})) \leqslant \sigma \, \text{wid}(\boldsymbol{x}), \tag{1.38}$$

then centered form (1.37) has an approximation property of order $n + 1$.

Higher-order enclosure methods can also be obtained by using a Taylor expansion. It is well-known that, if real multivariate function f is $(n + 1)$ times partially differentiable in Dom (f) with respect to all variables, then it can be written as

$$f(x) = P(x - c) + R(x), \tag{1.39}$$

where $c \in \boldsymbol{x} \subseteq$ Dom (f), $P(x)$ is an n degree multivariate Taylor polynomial that approximates f, and $R(x)$ is the error of this approximation continuous on $\boldsymbol{x}$ and, thus, is bounded. Let $\boldsymbol{R}$ denote an interval that encloses the approximation error. Then, for all $x \in \boldsymbol{x}$, it holds that

$$f(x) \in P(x - c) + \boldsymbol{R}, \tag{1.40}$$

and pair $(P, \boldsymbol{R})$ is called a *Taylor model* of f around c on Dom (f) [18]. Taylor models reduce the problem of bounding the range of a function to the problem of bounding the range of a polynomial. However, bounding the range of a polynomial is non-trivial [31], and it is rather expensive. If the nonlinear terms contribute less than the linear terms (e.g., when the boxes are narrow), the naive interval evaluation of $P(x)$ usually outperforms methods based on simple slopes [161], and a bicentered form for $P(x)$ may even result in the exact range. In this case, the resulting range of function f has approximation order $n + 1$, where n is the degree of $P(x)$. If all terms contribute strongly, the accuracy will be rather poor, probably poorer than centered forms with slopes, over sufficiently wide boxes may be even poorer than the naive interval evaluation [161]. A comprehensive overview of the literature on Taylor forms as well as a discussion on their uses and limits can be found, e.g., in [161].

Tight bounds for the range of a multivariate polynomial function over box $\boldsymbol{x}$ can be obtained by using the *Bernstein expansion* (see, e.g., [26, 58, 59, 63, 87, 130, 222]). Let I denote a *multiindex* defined as an m-tuple of non-negative indices $(i_1, \ldots, i_m) \in \mathbb{R}^m$. The comparisons and arithmetic operations on multiindices are understood component-wise [60]. For real vector $x \in \mathbb{R}^m$, its *multipower* $x^I = x_1^{i_1} \ldots x_m^{i_m}$. The m-fold sum $\sum_{I=0}^{N} = \sum_{i_1=0}^{n_1} \cdots \sum_{i_m=0}^{n_m}$, and the generalized binomial coefficient is defined by $\binom{N}{I} = \binom{n_1}{i_1} \ldots \binom{n_m}{i_m}$ [60, 279]. With this notation, an m-variate polynomial of degree N

$$P(x) = \sum_{I=0}^{N} a_I x^I, \tag{1.41}$$

can be represented over $\boldsymbol{x}$ in the following *Bernstein form*:

$$P(x) = \sum_{I=0}^{N} b_I B_{N,I}(x), \tag{1.42}$$

where $B_{N,I}$ is the Ith *Bernstein polynomial* of degree N,

$$B_{N,I}(x) = \binom{N}{I} \frac{(x - \underline{x})^I (\overline{x} - x)^{N-I}}{(\overline{x} - \underline{x})^N}, \tag{1.43}$$

and

$$b_I = \sum_{J=0}^{I} \frac{\binom{I}{J}}{\binom{N}{J}} (\overline{x} - \underline{x})^J \sum_{K=J}^{N} \binom{K}{J} \underline{x}^{K-J} a_K. \tag{1.44}$$

are the so-called *Bernstein coefficients*.

The Bernstein expansion has the property that the range of polynomial $P(x)$ over $\boldsymbol{x}$ is enclosed by the interval spanned by the minimum and maximum Bernstein coefficients, i.e.,

$$\mathrm{Rge}\,(P \mid \boldsymbol{x}) \subseteq \left[\min_{I=0,\ldots,N} b_I, \max_{I=0,\ldots,N} b_I \right]. \tag{1.45}$$

For sufficiently small boxes, the Bernstein form gives us the exact range [60].

The traditional approach to computing the minimum and maximum Bernstein coefficients (see, e.g., [57, 279]) has exponential complexity, which makes it infeasible for polynomials with many variables [60]. Smith [262] proposed a method for the representation and computation of the Bernstein coefficients, which is usually nearly linear with respect to the number of terms in a polynomial [60].

In an obvious way, Bernstein polynomials can be used to bound the range of rational functions $f = P/Q$, where P and Q are multivariate polynomials (see, e.g., [61, 154]). Without loss of generality, it can be assumed that P and Q have the same degree [154].

Theorem 1.12 (Narkawicz et al. [154]) *Let P and Q be real polynomials with Bernstein coefficients $b_I(P)$ and $b_I(Q)$, $I = 0, \ldots, N$, over box $\boldsymbol{x}$, respectively. Assume that all Bernstein coefficients $b_I(Q)$ have the same sign and are non-zero. Then, for $f = P/Q$ and for all $x \in \boldsymbol{x}$, it holds that*

$$\min_{I=0,\ldots,N} \frac{b_I(P)}{b_I(Q)} \leqslant f(x) \leqslant \max_{I=0,\ldots,N} \frac{b_I(P)}{b_I(Q)}. \tag{1.46}$$

In order to obtain better bounds, it is suggested in [62] to represent the rational function in form

$$f(x) = R(x) + \frac{P(x) - R(x)Q(x)}{Q(x)}, \tag{1.47}$$

where $R(x)$ is a linear approximation of f. Function $R(x)$ can be obtained, for example, by using the linear least square approximation of control points $(I/N, f(I/N))$ that are associated with the vertices of input box $\boldsymbol{x}$ (see, e.g., [62, 154]).

By combining Bernstein and Taylor forms (see, e.g., [92, 155, 156]), we can bound the range of a class of multidimensional functions. The idea of the combined *Taylor-Bernstein form* is the following: first, a function f is expanded in Taylor polynomial (1.39) of degree m over $\boldsymbol{x}$ (notice that f must be of class $\mathcal{C}^{m+1}$). Then, the Taylor polynomial is transformed to the Bernstein form. From (1.40) and (1.45), it follows that

$$\operatorname{Rge}(f \mid \boldsymbol{x}) \subseteq \left[\min_{I=0,\ldots,m} b_I, \max_{I=0,\ldots,m} b_I\right] + \boldsymbol{R}.$$

The Taylor-Bernstein form retains the approximation order of the Taylor form used to approximate f.

Bernstein polynomials are only one of the many polynomial functions that can be used in range bounding. For the one-dimensional case, Lin and Rokne [130] proposed expanding polynomial $P(x) = \sum_{i=0}^{n} a_i x^i$ over interval [0, 1] by using a more-general set of basis functions

$$\{A_j(x) \mid j = a, \ldots, b, \}, \ a < b, \tag{1.48}$$

where A_j, $j = a, \ldots, b$ fulfill the following conditions:

$$A_j(x) \geqslant 0, \ x \in [0, 1], \tag{1.49a}$$

$$\sum_{j=a}^{b} A_j(x) \equiv 1, \tag{1.49b}$$

$$x^i = \sum_{j=a}^{b} \mu_{ji} A_j(x),\ i = 0, \dots, n. \tag{1.49c}$$

From (1.49c), it follows that

$$P(x) = \sum_{i=0}^{n} a_i \sum_{j=a}^{b} \mu_{ji} A_j(x) = \sum_{j=a}^{b} \xi_j A_j(x), \tag{1.50}$$

where $\xi_j = \sum_{i=0}^{n} a_i \mu_{ji},\ j = 0, \dots, n.$

Lemma 1.13 *Let P be a real polynomial of degree n, and let $\{A_j(x) \mid j = a, \dots, b\}$ be a set of basis functions satisfying conditions (1.49). Then,*

$$\mathrm{Rge}\,(P \mid [0, 1]) \subseteq \left[\min_{j=a,\dots,b} \xi_j,\ \max_{j=a,\dots,b} \xi_j \right].$$

Proof Follows immediately from (1.50), (1.49a), and (1.49b). □

The assumption that input box $\boldsymbol{x} = [0, 1]$ is no restriction, since an arbitrary nonempty interval can be mapped onto interval [0, 1] by a linear transformation.

Theorem 1.14 (cf. Lin and Rokne [130]) *Let P be a real polynomial, and let $\{A_j(x) \mid j = a, \dots, b\}$ be a set of basis functions satisfying conditions (1.49). If*

$$\left| \sum_{i=0}^{n} a_i \mu_{ji} - P(x_j) \right| = \mathcal{O}(k^{-\alpha}),\ j = a, \dots, b$$

holds for some $x_j \in [0, 1]$, then

$$w\left(\left[\min_{j=a,\dots,b} \xi_j,\ \max_{j=a,\dots,b} \xi_j \right] \right) - w(\mathrm{Rge}\,(P \mid [0, 1])) = \mathcal{O}(k^{-\alpha}),$$

where k is a positive integer, and $\alpha > 0$.

The Bernstein and B-spline bases were described in [130]. The Chebyshev basis was discussed by Dzetkulič [44].

If continuous function f is monotonic on $\boldsymbol{x}$ in all its variables (i.e., $\frac{\partial f}{\partial x_i}$, $i = 1, \dots, n$, have a constant sign on $\boldsymbol{x}$), then $\mathrm{Rge}\,(f \mid \boldsymbol{x})$ is spanned by the values of f at the respective endpoints of input box $\boldsymbol{x}$:

$$\mathrm{Rge}\,(f \mid \boldsymbol{x}) = [f(x^-), f(x^+)], \tag{1.51}$$

where x^-, x^+ are defined by their components:

$$x_i^- = \begin{cases} \overline{x}_i & f \searrow \text{ on } \boldsymbol{x}, \\ \underline{x}_i & f \nearrow \text{ on } \boldsymbol{x}, \end{cases} \tag{1.52a}$$

$$x_i^+ = \begin{cases} \underline{x}_i & f \searrow \text{ on } \boldsymbol{x}, \\ \overline{x}_i & f \nearrow \text{ on } \boldsymbol{x}. \end{cases} \tag{1.52b}$$

For an n-variate function f that is of class C^1, monotonicity can be verified by computing the range of partial derivatives $\frac{\partial f}{\partial x_i}$, $i = 1, \ldots, n$ over $\boldsymbol{x}$. So,

$$x_i^- = \begin{cases} \underline{x}_i & \operatorname{Rge}\left(\frac{\partial f}{\partial x_i} \,\middle|\, \boldsymbol{x}\right) \geqslant 0, \\ \overline{x}_i & \operatorname{Rge}\left(\frac{\partial f}{\partial x_i} \,\middle|\, \boldsymbol{x}\right) \leqslant 0, \end{cases} \tag{1.53a}$$

$$x_i^+ = \begin{cases} \underline{x}_i & \operatorname{Rge}\left(\frac{\partial f}{\partial x_i} \,\middle|\, \boldsymbol{x}\right) \leqslant 0, \\ \overline{x}_i & \operatorname{Rge}\left(\frac{\partial f}{\partial x_i} \,\middle|\, \boldsymbol{x}\right) \geqslant 0. \end{cases} \tag{1.53b}$$

If function f is monotonic with respect to some variables only (i.e., it is *partial monotonic* [cf. [162, 173]]), then the following result can be used to compute the bounds for the range of f over $\boldsymbol{x}$.

Theorem 1.15 *Let f be a continuous function in n variables $(x_1, \ldots, x_n)$. Assume that variables x_i can be split into three disjoint groups (G_1, G_2, G_3) such that the monotonicity direction of f cannot be ascertained in G_1, f monotonically decreases on $\boldsymbol{x}$ for all x_i in G_2, and monotonically increases on $\boldsymbol{x}$ for all x_i in G_3. Then,*

$$\operatorname{Rge}(f \mid \boldsymbol{x}) = \left[\inf_{x \in \boldsymbol{x}'} f(x), \sup_{x \in \boldsymbol{x}''} f(x)\right], \tag{1.54}$$

where the entries of $\boldsymbol{x}'$ and $\boldsymbol{x}''$ are defined as follows:

$$\boldsymbol{x}_i' = \begin{cases} \boldsymbol{x}_i & x_i \in G_1, \\ \overline{x}_i & x_i \in G_2, \\ \underline{x}_i & x_i \in G_3, \end{cases} \tag{1.55a}$$

$$\boldsymbol{x}_i'' = \begin{cases} \boldsymbol{x}_i & x_i \in G_1, \\ \underline{x}_i & x_i \in G_2, \\ \overline{x}_i & x_i \in G_3. \end{cases} \tag{1.55b}$$

If necessary, input box $\boldsymbol{x}$ can be split to find sub-boxes on which f is guaranteed to be monotonic. Using splitting, we can obtain an arbitrary narrow enclosure for range [146], however, this technique is inefficient and, thus, not suitable for practical computation [160].

A result similar to the one described above can be obtained by using *Kaucher interval arithmetic* (see, e.g., [95, 96]), which is one of several modifications of classical interval arithmetic. In Kaucher arithmetic, the set of *proper* intervals is extended by the set of *improper* intervals $\mathbb{IR}^* = \{ [\underline{x}, \overline{x}] \mid \underline{x} \geqslant \overline{x} \}$. The elements of $\mathbb{KR} = \mathbb{IR} \cup \mathbb{IR}^* \cong \mathbb{R}^2$ are called *extended, directed*, or *generalized intervals*.

The conventional arithmetic operations, order relations, and other functions are isomorphically extended on $\mathbb{KR}$ [96]. Additionally, the `dual` operator is introduced, which swaps the endpoints of an interval, i.e., for $\boldsymbol{x} = [\underline{x}, \overline{x}] \in \mathbb{KR}$, $\text{dual}(\boldsymbol{x}) = [\overline{x}, \underline{x}]$.

The set of extended intervals possess group properties under addition and multiplication. Therefore, Kaucher arithmetic allows us to eliminate the dependency problem to a certain (rather small) extent in range computation.

Theorem 1.16 (Gardenes and Trepat [55]) *Let $f(x, y)$ be a rational function multi-incident on y, and there exists a splitting $y' = (y'_1, \ldots, y'_p)$, $y'' = (y''_1, \ldots, y''_q)$ of the incidences of y. Let $g(x, y', y'')$ correspond to the expression of f with explicit reference to the incidences of y, and $g(x, y', y'')$ is continuous on $\boldsymbol{x} \times \boldsymbol{y}' \times \boldsymbol{y}''$. Suppose that $g(x, y', y'')$ is unconditionally $\leqslant$-isotone for any component of y' and unconditionally $\leqslant$-antitone for any component of y'' on $\boldsymbol{x} \times \boldsymbol{y}' \times \boldsymbol{y}''$, then,*

- *if $f(x, y)$ is unconditionally $\leqslant$-isotone for y on $\boldsymbol{x} \times \boldsymbol{y}$, then*

$$\text{Rge}\,(f \mid (\boldsymbol{x}, \boldsymbol{y})) = \boldsymbol{G}(\boldsymbol{x}, \boldsymbol{y}', \text{dual}(\boldsymbol{y}'')) \subseteq \boldsymbol{F}(\boldsymbol{x}, \boldsymbol{y}), \tag{1.56}$$

- *if $f(x, y)$ is unconditionally $\leqslant$-antitone for y on $\boldsymbol{x} \times \boldsymbol{y}$, then*

$$\text{Rge}\,(f \mid (\boldsymbol{x}, \boldsymbol{y})) = \boldsymbol{G}(\boldsymbol{x}, \text{dual}(\boldsymbol{y}'), \boldsymbol{y}'') \subseteq \boldsymbol{F}(\boldsymbol{x}, \boldsymbol{y}), \tag{1.57}$$

where G and F are interval extensions of g and f, respectively.

In practice, the use of Theorem 1.16 is restricted to cases where the expression to be enclosed is totally monotonic. In the general case of non-monotonic functions, this methodology is not efficient.

In the following example, we compare some of the above-described methods for range bounding.

Example 1.17 Consider function $f : \mathbb{R} \to \mathbb{R}$ defined by expression

$$\mathsf{f}(x) = 1 - 5x + 1/3x^3.$$

The exact range of f over $\boldsymbol{x} = [0.5, 2.5]$ is $[-6.45356, -1.45833]$ (see Fig. 1.2). The naive evaluation yields the range $[-11.4583, 3.70833]$, which is about three times wider than the exact bounds. Table 1.1 presents the ranges obtained by using some of the above-described methods. The last column shows the relative overestimation.

To summarize: all of the above-mentioned approaches to range bounding can be used in solving parametric interval linear systems. However, to our best knowledge, only a few of them have been used so far to develop solution methods for PILS (see, e.g., [60, 173, 248]). Generally, two main approaches to solving PILS have been extensively studied over the past several years. The first one is based on Kaucher arithmetic, this was developed by Popova [173] and then studied by Popova (see, e.g.,

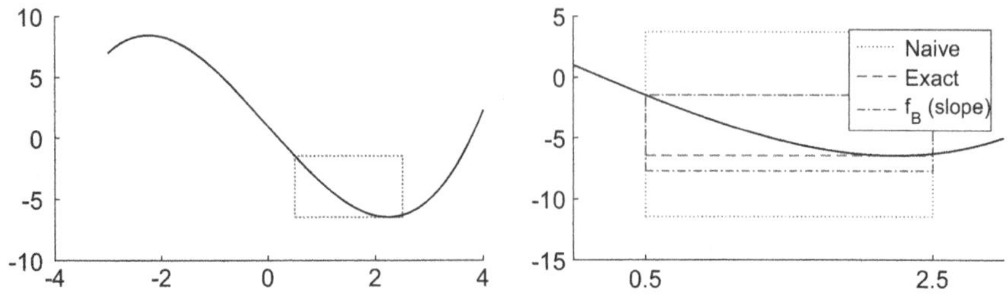

Fig. 1.2 Graph of function defined by arithmetic expression $\mathrm{f}(x) = 1 - 5x + 1/3x^3$ with exact range (left); exact range, naive bounds and bounds obtained by using bicentered form with slope (right)

Table 1.1 Comparison of selected range bounding methods for Example 1.17

Method	Range	Overestimation (%)
Slope form	$[-9.29167, -1.45833]$	36
Mean value form	$[-10.12500, -0.62500]$	47
Bicentered form (mean value)	$[-8.38175, -1.34742]$	29
Bicentered form (slope)	$[-7.7037, -1.45833]$	20

[178, 187]) and El-Owny [47], among others. The other one is based on alternative arithmetic, such as Hansen generalized arithmetic [72], Kolev's generalized intervals [106], and assorted variants of affine arithmetic (see, e.g., [33, 139]). The research conducted by the author of this book and presented in the following chapters is a part of the second approach. We use affine and revised affine forms in range computation. The main reason for this choice is that affine forms arise in the context of solving PILS in a natural way. Indeed, a parametric interval linear system with affine-linear dependencies can be viewed as a system with entries given as affine forms. Moreover, affine arithmetic reduces the so-called *wrapping effect*, which is related to overestimation caused by nested operations on dependent variables. Finally, affine arithmetic allows us to obtain a parametric solution (see Chap. 5) that can be used to solve various other problems involving interval parameters. Chapter 2 presents preliminary theories on affine forms along with some new results achieved by the author of this book over the past few years.

1.4 Rounded Interval Arithmetic

An important feature of interval arithmetic is that it can be implemented on a floating-point computer such that the resulting interval contains the result of the real-interval computation using equations (1.15) and directed rounding. The implementation of interval arithmetic described here is specific to the x86-32 instruction set architecture and is compliant with the IEEE 754–1985 standard for binary floating-point arithmetic.

Implemented in the hardware of most computers, the IEEE 754-1985 standard defines two basic binary floating-point formats called single precision (32-bit) and double precision (64-bit). Additionally, extended and extendable formats are permitted in the IEEE 754 standard. They are recommended to be used internally by processors or arithmetic co-processors to achieve higher precision in minimizing round-off errors than is provided by basic formats. The standard only specifies a minimum precision and exponent requirements for such formats. The x87 80-bit extended format is the most-commonly-implemented extended format that meets these requirements.

The IEEE 754-2008 standard defines the 64-bit double precision (`binary64`) format, which is of the main interest here, as having (from left to right in Windows systems) 1 sign bit, 11 bits for a biased exponent, and 53 bits for a normalized significand (also coefficient or mantissa) from which only 52 bits of a mantissa fraction are explicitly stored. This gives approximately 16 significant decimal digits of precision, $53/\log_2 10 \approx 15.955$.

The following 64-bit binary string

$$s\, e_{10}e_9 \ldots e_0\, f_{-1} f_{-2} \ldots f_{-52}$$

represents a floating-point number

$$x_F = (-1)^s \cdot \left(1 + \sum_{i=1}^{52} b_i 2^{-i}\right) \cdot 2^E, \quad E = e - 1023, \quad e = \sum_{i=0}^{10} e_i 2^i$$

from range $[-(1+(1-2^{-52}))2^{1043}, -2^{-1022}] \cup [2^{-1022}, (1+(1-2^{-52}))2^{1043}]$, unless fraction and exponent take some special values. The following four special cases can be distinguished:

- singed zeros, with $e = f = 0$,
- subnormal numbers (subnormals), with $e = 0$, $f \neq 0$, which fill the underflow gap around zero in floating-point arithmetic, i.e., they represent the floating-point numbers:

$$x_F = (-1)^s \cdot \sum_{i=1}^{52} b_i 2^{-i} \cdot 2^{-1022},$$

  from range $[-(1-2^{-52})2^{-1022}, -2^{1022-52}, 2^{1022-52}, (1-2^{-52})2^{-1022}]$,
- signed infinities, with $e =$#7ff and $f = 0$,
- NaNs, with $e =$#7ff and $f \neq 0$.

The finite precision of floating-point numbers results in the fact that only a finite subset of real numbers (a floating-point screen) can be represented on a computer, and true arithmetic operations cannot be precisely represented by floating-point operations. Let $\mathbb{F} \subset \mathbb{R}$ denote the finite set of floating-point numbers, i.e., machine representable numbers. Then, $\mathbb{IF}$ will denote the set of all machine representable intervals that are intervals with floating-point endpoints. Obviously, the result of an operation on representable intervals cannot always be represented exactly and, thus, must be rounded. In order to preserve the inclusion principle of interval arithmetic, the result $\boldsymbol{x} = [\underline{x}, \overline{x}] \in \mathbb{IR}$ of an interval operation must be *outwardly rounded*, that is, its lower endpoint $\underline{x}$ must be rounded down, and its upper endpoint $\overline{x}$ must be rounded up. Moreover, it is desirable that the rounding preserve as much information as possible, that is, the resulting machine representable interval $\boldsymbol{x}_F \in \mathbb{IF}$ is the tightest-such interval. The amount by which the resulting interval is widened is called the *roundout error*.

The IEEE 754 standard defines four rounding modes: to the nearest (default for all compilers), towards zero (truncation), towards positive infinity or ceiling (rounded up), and towards negative infinity or floor (rounded down). The last three modes are called the *directed rounding* modes. Let $\triangle : \mathbb{R} \to \mathbb{F} \cup \{+\infty\}$ and $\triangledown : \mathbb{R} \to \mathbb{F} \cup \{-\infty\}$ denote the directed rounding, respectively, towards positive and negative infinity. We say that they are *optimal* if the following equalities are satisfied:

$$\begin{aligned} \triangle x &= \inf\{x_F \in \mathbb{F} \mid x_F \geqslant x\}, \\ \triangledown x &= \sup\{x_F \in \mathbb{F} \mid x_F \leqslant x\}. \end{aligned}$$

The tightest machine representable interval containing $\boldsymbol{x} = [\underline{x}, \overline{x}] \in \mathbb{IR}$ is given by the *optimal outward* rounding

$$\Diamond \boldsymbol{x} = [\triangledown \underline{x}, \triangle \overline{x}] \in \mathbb{IF}. \tag{1.58}$$

Obviously, for $\boldsymbol{x} \in \mathbb{IR}$, it holds that $\boldsymbol{x} \subseteq \Diamond \boldsymbol{x}$ with equality if the endpoints of $\boldsymbol{x}$ are machine representable numbers. Moreover, optimal outward rounding are inclusion monotonic, i.e. for $\boldsymbol{x}, \boldsymbol{y} \in \mathbb{IR}$, $\boldsymbol{x} \subseteq \boldsymbol{y}$ implies $\Diamond \boldsymbol{x} \subseteq \Diamond \boldsymbol{y}$.

With an optimal outward rounding, interval arithmetic operations $\odot$ and elementary functions $\varphi^{\Diamond}$ can be defined on a computer as (cf. [160]):

$$\begin{aligned} \boldsymbol{x} \odot \boldsymbol{y} &= \Diamond(\boldsymbol{x} \bullet \boldsymbol{y}) \supseteq \boldsymbol{x} \bullet \boldsymbol{y}, \quad \text{for } \bullet \in \{+, -, *, /\}, \\ \varphi^{\Diamond}(\boldsymbol{x}) &= \Diamond\varphi(\boldsymbol{x}) \ni \tilde{x}, \quad \text{for all } \tilde{x} \in \boldsymbol{x}. \end{aligned}$$

Changing the rounding mode is an expensive operation. However, relations $\triangledown x = -\triangle(-x)$ and $\triangle x = -\triangledown(-x)$ usually hold for $x \in \mathbb{R}$; thus, interval arithmetic operations can be implemented by setting only one rounding mode per operation. It is possible as well to simulate interval operations without changing the rounding mode by using, e.g., the proposed in [233] algorithms for computing the predecessor and successor of a floating-point number.

The directed rounding that are necessary for outward rounding are implemented in the hardware of IEEE-compliant floating-point units (FPUs). They are optimal for elementary arithmetic operations (addition, subtraction, multiplication, division, and square root). The IEEE 754 standard requires that the *relative error* of the result of these operations in rounding to nearest mode must not be greater than $\frac{1}{2}$ulp (in the directed rounding mode, the relative error should not be greater than 1ulp), where ulp stands for *unit in the last place* or *unit of the least precision*). Let $\mathrm{fl} : \mathbb{R} \to \mathbb{F}$ denote rounding to nearest according to IEEE-754. Then, for $\bullet \in \{+.-, *, /\}$ and $x, y \in \mathbb{F}$,

$$\mathrm{fl}(x \bullet y) = (x \bullet y)(1 + \delta) + \mu, \tag{1.59}$$

where $|\delta| \leqslant \mathbf{u} = \frac{1}{2}$ulp and $|\mu| \leqslant \frac{1}{2}\eta$ (η denotes the smallest positive subnormal floating-point number), and at least one of δ, μ is zero (for addition and subtraction μ is always zero), provided $\mathrm{fl}(x \bullet y)$ is finite [233]. An important property of fl is its monotonicity: for all $x, y \in \mathbb{R}$ it holds that

$$x \leqslant y \Rightarrow \mathrm{fl}(x) \leqslant \mathrm{fl}(y). \tag{1.60}$$

In particular, this means that the rounding cannot "jump" over a FP number [233], i.e., if $x \in \mathbb{R}$, $x_F \in \mathbb{F}$ $x \leqslant x_F$ then $\mathrm{fl}(x) \leqslant x_F$.

For the basic transcendental functions (e.g., logarithm, exponential, or trigonometric function) it is required in prestigious numerical libraries that the relative error is between $\frac{1}{2}$ulp and around 1ulp of the mathematically exact result. This means that the relative error of $\Diamond\varphi(\boldsymbol{x})$ should not be greater than 2ulps. If optimal rounding is not available, a suboptimal (artificial) outward rounding can be defined using auxiliary functions (see, e.g., [160, 233]). For $x, y \in \mathbb{F}$ and finite $z := \mathrm{fl}(x \bullet y)$, the monotonicity property (1.60) implies

$$x \bullet y \in [\mathrm{fl}(\mathrm{fl}(z - \mathrm{fl}(2\mathbf{u}|z|)) - \eta), \mathrm{fl}(\mathrm{fl}(z + \mathrm{fl}(2\mathbf{u}|z|)) + \eta)].$$

The above formula remains valid if $x \bullet y$ is replaced by $y = \varphi(x)$, as long as z is the correct rounding of y [233].

1.4.1 Setting Rounding Mode

On systems based on the x86-32 architecture, the rounding settings are stored in the 16-bit Floating-Point Control Word (FCW) Register of the Floating-Point Unit. The rounding flags (RC bits) determine the method of rounding that the FPU uses (see Table 1.2).

The assembly language procedures given below (RoundUp and RoundDown set the RC bits according to their names.

```
int CRoundDown = 0x0400; // 0000 01 0000000000
int CRoundUp = 0x0800; // 0000 10 0000000000
int CRoundUp = 0x0800; // 0000 10 0000000000
int CRoundNear = 0xF3FF; // 1111 00 1111111111
```

Table 1.2 Rounding flags settings

RC	Mode	The result of rounding for $x \in \mathbb{R}$
00	Round to nearest	$x_{RN} = x(1+\delta),\ \lvert\delta\rvert \leqslant 2^{-p}$
01	Round down	$x_{RD} = \max\{x_F \in \mathbb{F} \mid x_F \leqslant x\}$
10	Round up	$x_{RU} = \min\{x_F \in \mathbb{F} \mid x_F \geqslant x\}$

```
void RoundDown()
{ _ _asm{ fstcw oldcw
    fwait
    mov eax, oldcw
    and eax, CRoundNear
    or eax, CRoundDown
    mov newcw, eax
    fldcw newcw }
 }
```

```
void RoundUp()
{ _ _asm{ fstcw oldcw
    fwait
    mov eax, oldcw
    and eax, CRoundNear
    or eax, CRoundUp
    mov newcw, eax
    fldcw newcw }
}
```

Hardware rounding is implemented by overloading the arithmetic operators as follows:

```
interval operator + (interval x, interval y)
{
    double inf, sup;
    RoundUp();
    inf = -x.inf - y.inf;
    sup = x.sup + y.sup;
    return interval(-inf, sup);
}
```

Chapter 2
Alternative Arithmetic

Over the years, a lot of effort has been put into the development of self-validated computational (SVC) models that will be able to overcome the "memoryless nature" of interval arithmetic, i.e., to take into account the dependencies between variables involved in a computation and/or reduce the so-called wrapping effect. This effort has resulted in several such models worth mentioning: *ellipsoid calculus* [27, 269], *constrained interval arithmetic* [131], *Hansen's generalized interval arithmetic* [72], *affine arithmetic* [33], *reduced affine arithmetic* [139], and *revised affine arithmetic* [271].

In this chapter, we present the main concepts of the affine and revised affine arithmetic. We put the main focus on the latter, since it is our main tool for developing methods for solving parametric interval systems.

2.1 Affine Arithmetic

Affine arithmetic (AA) [28, 33–35] was first introduced by Comba and Stolfi [28] as a new self-validated model for numerical computation, even though a similar model was developed in the early eighties by Hansen [72] under the name *generalized intervals* (GI). Both GI and AA were designed to eliminate the main weakness of (standard) interval arithmetic [146], that is, the tendency to produce intervals that are often much wider than the true range of the corresponding quantities, especially in long computation chains (see Fig. 2.1).

Example 2.1 Consider the following rational function

$$g(x) = \frac{x^2 - x + 1/2}{x^2 + 1/2}. \tag{2.1}$$

I. Skalna, *Parametric Interval Algebraic Systems*, Studies in Computational Intelligence 766, https://doi.org/10.1007/978-3-319-75187-0_2

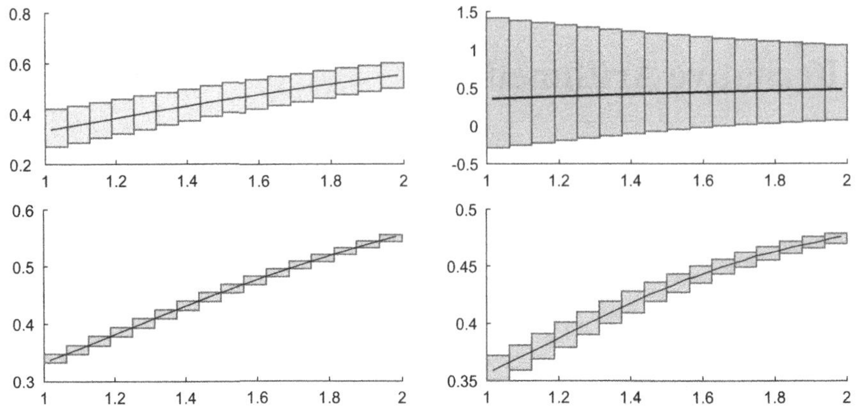

Fig. 2.1 Overestimation of IA estimates: graph of function $g(x)$ on interval $x \in [1, 2]$ (black solid line) and values of $\boldsymbol{g}(x)$ for 16 consecutive equal subintervals (top left); graph of function $h(x) = g^3(x)$ and values of $\boldsymbol{h}(x)$ for same data (top right); same computation with affine arithmetic: $\boldsymbol{g}(x)$ (bottom left), $\boldsymbol{h}(x)$ (bottom right)

Figure 2.1 shows the graphs of functions $g(x)$ and $h(x) = g^3(x)$, for $x \in [1, 2]$, as well as the results of (naive) interval evaluation of $\boldsymbol{g}(x)$ and $\boldsymbol{h}(x)$ for 16 consecutive equal subintervals of $[1, 2]$. Although the iterates g^n converge to a constant function (≈ 0.397), the interval ranges computed using IA diverge, since the overestimation factors of the individual steps tend to get multiplied.

Affine arithmetic, like standard interval arithmetic, produces guaranteed enclosures for the computed quantities, taking into account any uncertainties in the input data as well as all internal truncation and round-off errors [34]. In addition, affine arithmetic keeps track of first-order correlations,[1] thus, it eliminates the dependency problem of interval arithmetic to some extent. Thanks to this extra feature, AA is able to provide much-tighter bounds for the computed quantities than standard IA [33] (see Fig. 2.1). Unfortunately, this advantage comes at the expense of higher computational effort.

In AA [28, 33–35], unknown ideal quantity $\tilde{x}$ is represented by affine form

$$\hat{x} \triangleq x_0 + x_1\varepsilon_1 + \cdots + x_n\varepsilon_n \tag{2.2}$$

which is a degree 1 polynomial. We write $\tilde{x} \in [\hat{x}]$ to symbolically denote that $\tilde{x}$ is represented by $\hat{x}$. *Central value* x_0 and *partial deviations* x_i are real numbers (in computer implementation, they are finite floating-point numbers), and *noise symbols* (sometimes called *noise variables* [34] or *dummy variables* [235, 263]) ε_i are unknown but assumed to vary independently within interval $\boldsymbol{U} = [-1, +1]$.

[1] Hansens's generalized interval arithmetic records only the correlations between output quantities z_i and inputs x_j, but not among the inputs nor the outputs between the quantities involved in a computation [33].

Each noise symbol stands for an independent component of the total uncertainty of unknown ideal quantity $\tilde{x}$. The *Fundamental Invariant of Affine Arithmetic* (FIAA), which formalizes the semantic of affine forms, states that

> *At any stable instant in AA computation, there is a single assignment of values from* $\boldsymbol{U}$ *to each of the noise variables in use at the times that makes the value of every affine form equal to the value of the corresponding quantity in the ideal computation* [33].

A *stable instant* is the time when the algorithm is not performing an AA operation [33].

Affine forms sharing the same noise symbols are partially correlated through them, and coefficients x_i and y_i determine the magnitude of this correlation. The signs of the coefficient are themselves not relevant, but the relative sign of x_i and y_i defines the direction of the correlation [33].

Every affine form $\hat{x} = x_0 + \sum_{i=1}^{n} x_i \varepsilon_i$ implies the bounds for corresponding ideal quantity $\tilde{x}$, i.e., interval

$$\tilde{x} \in \left[\hat{x}\right] = [x_0 - \|x\|_1, x_0 + \|x\|_1],$$

where $\|x\|_1 = \sum_{i=1}^{n} |x_i|$ is called the *total deviation* [34] or the radius of $\hat{x}$. Range $[\hat{x}]$ is the smallest interval that contains all possible values of $\hat{x}$ and, thus, also unknown ideal quantity $\tilde{x}$, assuming that each ε_i varies independently within its domain. Obviously, this conversion discards all information about the dependence between the variables involved in a computation.

Conversely, if ideal but unknown quantity $\tilde{x}$ belongs to some interval $\boldsymbol{x} = \left[\underline{x}, \overline{x}\right]$, then $\tilde{x}$ can be represented by the affine form

$$\hat{x} = x^c + x^\Delta \varepsilon_k,$$

where ε_k is a new noise symbol (different from other noise symbols) unless some information about the dependence between $\tilde{x}$ and the other variables involved in a computation is available. In floating-point computations, x^Δ should be large enough to compensate for the rounding error made in computing x^c [33].

Let $\hat{x}$, $\hat{y}$ be two affine forms. Then, each point $(\tilde{x}, \tilde{y})$, where $\tilde{x} \in [\hat{x}]$ and $\tilde{y} \in [\hat{y}]$, can be expressed as

$$(\tilde{x}, \tilde{y}) = (x_0, y_0) + \sum_{i=1}^{n} \varepsilon_i (x_i, y_i). \tag{2.3}$$

Hence, all possible points with coordinates represented by affine forms $\hat{x}$ and $\hat{y}$ (assuming that each ε_i vary independently on $\boldsymbol{U}$) lie in a convex polygon (*zonotope*), which is called a *joint range* (Fig. 2.2) and denoted by $\langle\hat{x}, \hat{y}\rangle$. A joint range is centrally symmetric around point (x_0, y_0), compact and convex.

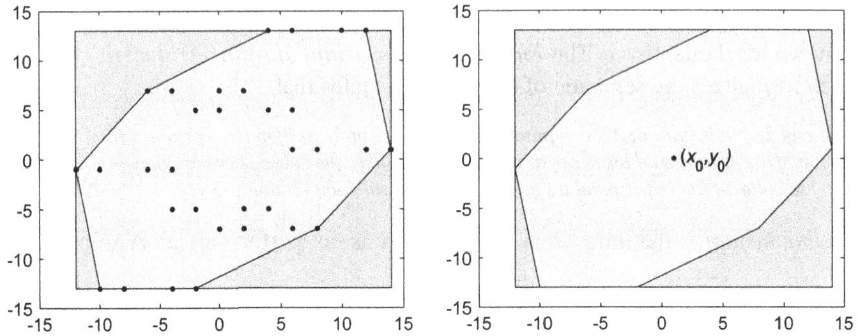

Fig. 2.2 Joint range of affine forms $\hat{x} = 1 - \varepsilon_1 + 3\varepsilon_2 - 5\varepsilon_3 - \varepsilon_4 - 3\varepsilon_5$, $\hat{y} = 6\varepsilon_1 - 3\varepsilon_3 - 4\varepsilon_5$ (light gray region) and box $[\hat{x}] \times [\hat{y}]$ (dark gray region); points (black dots) obtained for $\varepsilon_1, \ldots, \varepsilon_5 \in \{-1, 1\}$ (left); central point (right)

2.1.1 Arithmetic Operations on Affine Forms

Elementary operations on affine forms are redefined so that they result in affine forms. Affine-linear operations result in a straightforward manner in affine forms. Except for round-off errors and overflows (which must be handled through appropriate implementation), a resulting affine form contains all of the information about the result that can be deduced from the affine operands. In particular, given two arbitrary affine forms $\hat{x}$, $\hat{y}$ and $\alpha, \beta, \gamma \in \mathbb{R}$, the following equality holds true:

$$\alpha\hat{x} + \beta\hat{y} + \gamma = \alpha x_0 + \beta y_0 + \gamma + e^T(\alpha x + \beta y), \tag{2.4}$$

where $e = (\varepsilon_1, \ldots, \varepsilon_n) \in \boldsymbol{U}^n$ is a vector of the noise symbols. It is worth noting that the formula above guarantees that $\hat{x} - \hat{x} = 0$, which is not possible in interval arithmetic unless $x^{\Delta} = 0$. This property is achieved thanks to the fact that the affine forms being subtracted share the same noise symbols with the same coefficients, therefore, they are treated as the same quantity and not two different quantities having the same range.

Non-affine operations on affine forms cannot be computed in the same straightforward manner, since they usually result in nonlinear functions in the noise symbols. Without loss of generality, we can restrict our considerations to the non-affine operations that are functions of two variables. Thus, given two affine forms $\hat{x}$ and $\hat{y}$, we must compute an affine form $\hat{z}$, which represents $z = f\left(\hat{x}, \hat{y}\right)$. Clearly, $z = f\left(\hat{x}, \hat{y}\right)$ can be considered as a function of the noise symbols:

$$z = f\left(x_0 + e^T x,\ y_0 + e^T y\right) = f^*(e). \tag{2.5}$$

If f^* is not affine, then we must select affine function

$$f^a\left(\hat{x},\hat{y}\right) = f^a\left(e\right) = z_0 + z_1\varepsilon_1 + \ldots + z_n\varepsilon_n \tag{2.6}$$

that approximates f^* on $\langle\hat{x},\hat{y}\rangle$ reasonably well. There are $n+1$ degrees of freedom for the choice of affine approximation f^a. However, it is desirable that resulting affine form $\hat{z}$ is consistent with $\hat{x}$ and $\hat{y}$ and that it preserves as much information provided by $\hat{x}$ and $\hat{y}$ as possible. More specifically, $\hat{z}$ should preserve as much constraint between $\hat{x}$, $\hat{y}$, and $\hat{z}$ as possible without imposing any other constraints that cannot be deduced from the available data [33]. Still, the task is challenging, since different approximations may be more-accurate for specific purposes.

Generally, two basic approaches to compute affine approximation f^a are most-frequently used: the *minimum-error* and *minimum-range* approximation. The first one is a subject of *Chebyshev approximation theory*. It minimizes the maximum absolute error, however, the implied range can be quite wide, often wider than the range obtained using interval arithmetic. In order to minimize the range, we can use the second approach, it is called the *minimum-range* (or briefly, the *min-range*) approximation. Min-range approximation preserves less information but yields no overestimation, and thus can never be worse than interval arithmetic. The choice of which affine approximation to use strongly depends on the problem to be solved. For example, for some univariate functions such as the square, square root, or exponential functions, the min-range approximation is recommended since it guarantees that the implied range is positive (which is expected, as those functions are never negative). The range optimality is also needed for bounding the range of rational functions. In this case, a narrower denominator is less likely to contain zero.

Often, in order to maintain the simplicity and efficiency of algorithms, we restrict ourselves to approximations that are themselves linear combinations of input forms $\hat{x}$ and $\hat{y}$, that is:

$$f^a(\hat{x},\hat{y}) = \alpha\hat{x} + \beta\hat{y} + \gamma, \tag{2.7}$$

where $\alpha, \beta, \gamma \in \mathbb{R}$. Then, there are only three degrees of freedom for the choice of an affine approximation f^a, but we still have an infinite number of possibilities.

Once the affine approximation is selected, we must add an extra term (*error term*) to f^a, which represents the approximation error (in rigorous computation, it also includes round-off errors):

$$d^*(\hat{x},\hat{y}) = d^*\left(e\right) = f^*\left(e\right) - f^a\left(e\right). \tag{2.8}$$

Thus, affine form $\hat{z}$ that describes $f(\hat{x},\hat{y})$ is given by

$$\hat{z} = f^a\left(e\right) + z_{n+1}\varepsilon_{n+1} = z_0 + e^T z + z_{n+1}\varepsilon_{n+1}, \tag{2.9a}$$

$$z_0 = \alpha x_0 + \beta y_0 + \gamma, \tag{2.9b}$$

$$z = \alpha x + \beta y, \tag{2.9c}$$

where ε_{n+1} is a new noise symbol and z_{n+1} satisfies the inequality:

$$|z_{n+1}| \geqslant \max_{e \in U^n}\{|d^*(e)|\}.$$

Obviously, substituting $z_{n+1}\varepsilon_{n+1}$ for $d^*(e)$ generates a loss of some information, since noise symbol ε_{n+1} is assumed to be independent from the other noise symbols (whereas it is, in fact, a non-linear function of them). Any subsequent operation with $\hat{z}$ as an operand will not be aware of the dependence between ε_{n+1} and e, and they may subsequently return a less precise affine form. This is, however, a trade-off between the loss of accuracy and the simplicity of computation [33].

2.1.2 *Affine Approximations of Basic Univariate Nonlinear Functions*

For univariate nonlinear function $f : \mathbb{R} \to \mathbb{R}$, it can be shown that the (best) Chebyshev minimum-error approximation is of the following form [33]:

$$f^a(\hat{x}) = \alpha\hat{x} + \gamma, \tag{2.10}$$

where $\alpha, \gamma \in \mathbb{R}$.

Theorem 2.2 (Stolfi and de Figueiredo [33]) *Let f be a bounded and twice-differentiable real-valued function defined on interval $\boldsymbol{x} = [a, b]$ such that f'' does not change the sign inside $\boldsymbol{x}$, and let $f^a = \alpha x + \gamma$ be its Chebyshev minimum-error affine approximation in $\boldsymbol{x}$. Then,*

$$\alpha = (f(b) - f(a))/(b - a), \tag{2.11a}$$
$$\gamma = (f(u) + r(u))/2 - \alpha u, \tag{2.11b}$$
$$\delta = |f(u) - r(u)|/2, \tag{2.11c}$$

where $r(x)$ is the secant line passing through points $(a, f(a))$ and $(b, f(b))$, u satisfies the equality $f'(u) = \alpha$, and δ is the maximum absolute error.

We will now prove that if f satisfies the conditions of Theorem 2.2, then the Chebyshev approximation overestimates only one of the endpoints of the exact range by 2δ, whereas the other endpoint is computed exactly.

Corollary 2.3 *Let $\hat{x}$ be an affine form and let f be a monotone function that satisfies the conditions of Theorem 2.2 in $[\hat{x}]$. If $f'' > 0$ ($f'' < 0$) in $[\hat{x}]$, then range of Chebyshev minimum-error approximation $\hat{z}$ of f over $[\hat{x}]$ satisfies $[\hat{z}] = \mathrm{Rge}\left(f \mid [\hat{x}]\right) + [-2\delta, 0]$ ($[\hat{z}] = \mathrm{Rge}\left(f \mid [\hat{x}]\right) + [0, 2\delta]$).*

Proof Put $[a, b] = [\hat{x}]$ and assume that $f'' > 0$. Then, $\delta = (r(u) - f(u))/2$. The secant line passing through points $(a, f(a))$ and $(b, f(b))$ is given by

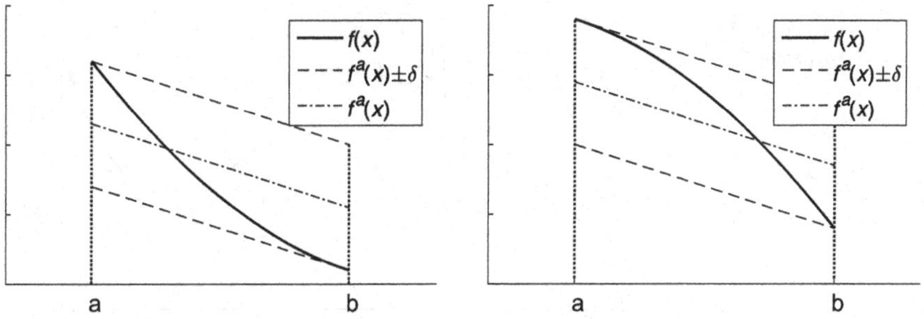

Fig. 2.3 Geometrical illustration of minimum range approximation of univariate function f over interval $[a, b]$; $f'' > 0$, $f' < 0$ (left); $f'' < 0$, $f' < 0$ (right)

$$r(x) = \alpha x + \frac{bf(a) - af(b)}{b - a}. \tag{2.12}$$

If f is monotone increasing, then the true range of f over $[a, b]$ is $[f(a), f(b)]$, and from (2.10) and Theorem 2.2, we get that the lower and upper bounds of $\hat{z}$ are as follows:

$$\begin{aligned}\inf([\hat{z}]) &= \alpha a + \gamma - \delta = \\ &= \alpha a + (f(u) + r(u))/2 - \alpha u - (r(u) - f(u))/2 = \\ &= \alpha a - \alpha u + f(u).\end{aligned}$$

$$\begin{aligned}\sup([\hat{z}]) &= \alpha b + \gamma + \delta = \\ &= \alpha b + (f(u) + r(u))/2 - \alpha u + (r(u) - f(u))/2 = \\ &= \alpha b - \alpha u + r(u) = \alpha b + \frac{bf(a) - af(b)}{b - a} = f(b).\end{aligned}$$

So, the upper bound of $[\hat{z}]$ equals the upper bound of the true range, and the lower bound of $[\hat{z}]$ overestimates the lower bound of the true range by

$$f(a) - \alpha a + \alpha u - f(u) = r(u) - f(u) = 2\delta.$$

The proof of the remaining cases is analogous. □

In the next proposition, we provide the basis for an algorithm for computing the min-range approximation of the univariate functions that are bounded monotone and concave or convex in $\boldsymbol{x} \in \mathbb{IR}$ (see Fig. 2.3).

Proposition 2.4 *Let f be a bounded monotone and twice-differentiable function in $\boldsymbol{x} = [a, b]$ such that f'' does not change the sign inside $\boldsymbol{x}$. Then, the min-range approximation for $z = f(x)$ in $\boldsymbol{x}$ can be defined as*

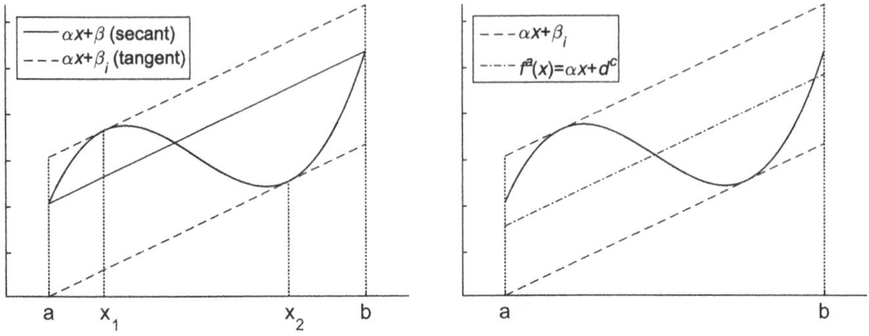

Fig. 2.4 Geometric illustration of affine-linear approximation of non-monotonic smooth function f over interval $[a, b]$

$$f^a(x) = \alpha x + \gamma, \tag{2.13}$$

where

$$\alpha = \begin{cases} f'(a), & f'' \cdot f' > 0, \\ f'(b), & f'' \cdot f' < 0, \end{cases} \tag{2.14}$$

$$\gamma = \begin{cases} \mathrm{mid}([f(a) - \alpha a, f(b) - \alpha b]), & f' > 0, \\ \mathrm{mid}([f(b) - \alpha b, f(a) - \alpha a]), & f' < 0. \end{cases} \tag{2.15}$$

The maximum absolute error is

$$\delta = \begin{cases} \mathrm{rad}([f(a) - \alpha a, f(b) - \alpha b]), & f' > 0, \\ \mathrm{rad}([f(b) - \alpha b, f(a) - \alpha a]), & f' < 0. \end{cases} \tag{2.16}$$

Proof The proof will be limited to the case when $f'' > 0$ and $f' > 0$. All of the remaining cases can be proven similarly.

If $f'' > 0$ and $f' > 0$ in $[a, b]$, then for each $x \in [a, b]$, $f(x) \in [f(a), f(b)]$. Now, it is enough to show that

$$\mathrm{Rge}\left(f^a(x) + \delta\varepsilon_k \,\middle|\, ([a, b], [-1, 1])\right) = [f(a), f(b)].$$

Put $\boldsymbol{y} = \mathrm{Rge}\,(f^a(x) + \delta\varepsilon_k \mid ([a, b], [-1, 1]))$. Then, by (2.14)–(2.16), we have

$$\boldsymbol{y} = f'(a)[a, b] + [f(a) - f'(a)a, f(b) - f'(a)b] = [f(a), f(b)].$$

□

If an univariate function is non-monotonic (cf. Fig. 2.4), it can be approximated over interval $[a, b]$ by the following affine form (cf. Kolev [103]):

$$\hat{z} = \alpha\hat{x} + d^c + d^\Delta\varepsilon_k, \tag{2.17}$$

where noise symbol ε_k must be distinct from all existing noise symbols, whereas real parameter α and interval $\boldsymbol{d} = [d^c - d^\Delta, d^c + d^\Delta]$ can be computed by using the following procedure (see [103]) shown in Algorithm 1.

Algorithm 1 Affine approximation of non-monotonic smooth univariate functions over interval

---

Input: A nonlinear non-monotonic smooth univariate function f,
 a closed interval $[a, b] \subset D_f$
Output: An affine approximation for function f over $[a, b]$:
 $\hat{z} = \alpha\hat{x} + d^c + d^\Delta \varepsilon_k$

// Compute the slope and intercept of a secant line
// passing through points $(a, f(a))$ and $(b, f(b))$
$\alpha = (f(b) - f(a))/(b - a)$ // slope
$\beta = f(a) - \alpha a = f(b) - \alpha b$ // intercept
Find in $[a, b]$ all zeros $x_1, \ldots, x_K$ of function $g(x) = f'(x) - \alpha$
Compute $\beta_i = f(x_i) - \alpha x_i, i = 1, \ldots, K$
Set $\boldsymbol{d} = [\underline{d}, \overline{d}]$, where
$\underline{d} = \min\{\beta, \min\{\beta_i, i = 1, \ldots, K\}\}$
$\overline{d} = \max\{\beta, \max\{\beta_i, i = 1, \ldots, K\}\}$
return $\hat{z} = \alpha\hat{x} + d^c + d^\Delta \varepsilon_k$

As we have seen, each non-affine operation and nonlinear function generate a new noise symbol. In rigorous computation, this extra term is generated by each operation on affine forms. So, the length of the affine forms gradually increases during the course of a computation. This causes that affine arithmetic is computationally very costly, which in turn limits its use in practical applications.

Some optimization techniques for AA programs have been described in [33]. One of them is to "condense" affine forms, i.e., to replace two or more terms $z_i\varepsilon_i$, $z_j\varepsilon_j$, ... by a single term $z_k\varepsilon_k$, where $z_k = |z_i| + |z_j| + \ldots$, and ε_k is a new noise symbol (distinct from the other noise symbols). In order to preserve as much information as possible, only "internal" noise symbols (i.e., noise symbols that were created during a computation) should be condensed [33]. In [139], the idea of "condensation" resulted in AF1 forms, also called *reduced affine forms*. The length of a reduced affine form is established at the beginning of a computation and remains unchanged thanks to the use of the so-called *accumulative error*, which represents all errors introduced by performing nonlinear operations. A similar idea lies in Kolev's *generalized intervals* (G-intervals) [104]. The "combination" of these two approaches is reflected in *revised affine forms* [271], which are described in the next section along with some new results we have obtained in recent years.

2.2 Revised Affine Arithmetic

Revised affine arithmetic (RAA) [271] is somewhat similar to reduced affine arithmetic [139] and Kolev's generalized intervals [104]. They all have a similar form and the same geometry.[2] A revised affine form is defined by

$$\hat{x} \triangleq x_0 + e^T x + x_r[-1, 1], \tag{2.18}$$

i.e., it is the sum of an affine form of length n and interval $x_r[-1, 1]$ ($x_r \geqslant 0$). Length n is equal to the number of initial ("external") uncertainties (due to measurement errors, numerical approximation, or indeterminacy) and remains unchanged during the same computation. Interval $x_r[-1, 1]$, which is referred to as the *accumulative error*, represents "internal" noise symbols, which are due to errors introduced by performing non-affine operations; in rigorous computations, it is also used to accumulate rounding errors. The use of the accumulative error does not influence the affine information since all of the internal noise symbols are, by assumption (made to simplify the computation), completely independent from "external" noise symbols and each other. The conversion between affine forms and intervals is the same as in the case of affine forms, i.e.,

$$x_0 + e^T x + x_r[-1, 1] \Rightarrow [x_0 - (\|x\|_1 + x_r), x_0 + (\|x\|_1 + x_r)],$$
$$x^c + x^\Delta \varepsilon_k \Leftarrow \boldsymbol{x} = [x^c - x^\Delta, x^c + x^\Delta],$$

where ε_k is a new noise symbol, independent from other noise symbols already used in a computation.

The general concepts of affine forms presented in the previous section apply to revised affine forms. This means, inter alia, that affine operations are straightforward, however, we must be aware that accumulative errors can never be canceled. So, given two revised affine forms $\hat{x}$ and $\hat{y}$ as well as some real scalars α, β, γ, their linear combination is given by

$$\hat{z} = \alpha\hat{x} + \beta\hat{y} + \gamma = z_0 + e^T z + z_r[-1, 1], \tag{2.19a}$$
$$z_0 = \alpha x_0 + \beta y_0 + \gamma, \tag{2.19b}$$
$$z = \alpha x + \beta y, \tag{2.19c}$$
$$z_r = |\alpha| x_r + |\beta| y_r. \tag{2.19d}$$

Non-affine operations, such as the multiplication or division of revised affine forms, must be handled in a similar way as non-affine operations on affine forms. This means that, from among an infinite number of revised affine forms, we must pick one that will approximate the result of an operation in the best-possible way. In the next section, we will pay more attention to the multiplication of affine forms,

[2] In Hansen's generalized intervals, the joint range may be a non-convex polygon.

since it constitutes the main bottleneck in affine computation. Not surprisingly, it has gained a lot of interest [33, 104, 110, 140, 144, 235, 265, 271, 280] since the first works on generalized intervals and on affine arithmetic appeared.

2.2.1 Multiplication

The majority of the multiplication formulae presented below were developed by us (Skalna and Hladík [258]) based on the arithmetic of standard affine forms and of generalized intervals.

The product of two revised affine forms

$$\hat{x} = x_0 + e^T x + x_r[-1, 1],$$
$$\hat{y} = y_0 + e^T y + y_r[-1, 1],$$

is a quadratic form on the noise symbols, which can be written as

$$f^*(e) = \hat{x} \cdot \hat{y} = \left(x_0 + e^T x + x_r[-1, 1]\right) \cdot \left(y_0 + e^T y + y_r[-1, 1]\right) = \tag{2.20a}$$
$$= x_0 y_0 + e^T (x_0 y + y_0 x) + \tag{2.20b}$$
$$+ x_0 y_r[-1, 1] + y_0 x_r[-1, 1] + x_r y_r[-1, 1] + \tag{2.20c}$$
$$+ y_r e^T x[-1, 1] + x_r e^T y[-1, 1] + \left(e^T x\right)\left(e^T y\right). \tag{2.20d}$$

Most of the existing approaches (see, e.g., [104, 140, 271]) use the (2.20b) part as an affine approximation, bound the error ((2.20c) + (2.20d)) of this approximation, and then add these bounds to (2.20b).

Different bounds on the approximation error yield different multiplication formulae (with respect to the accuracy and time complexity). Using the trivial range estimate [33, 140], revised affine form $\hat{z}$ that describes the product is defined as follows:

$$\hat{z} = x_0 y_0 + e^T (y_0 x + x_0 y) + z_r[-1, 1], \tag{2.21a}$$
$$z_r = |y_0| x_r + |x_0| y_r + (\|x\|_1 + x_r) \cdot (\|y\|_1 + y_r). \tag{2.21b}$$

The number of multiplications in (2.21) is $2n + 4$, and the number of additions is $3n + 2$, which gives asymptotic time complexity $\mathcal{O}(n)$, where n is the number of noise symbols. Multiplication formula (2.21) will be referred to as *trivial multiplication*.

Proposition 2.5 (cf. de Figueiredo and Stolfi [33], Messine [140]) *The affine multiplication defined by (2.21) is an inclusion function of $f(x, y) = x \cdot y$ over joint range $\langle \hat{x}, \hat{y} \rangle$.*

Tighter bounds on the quadratic term in (2.20d) can be obtained from the results presented in [106, 271]. Revised affine form $\hat{z}$ is defined in this case as follows:

$$\hat{z} = x_0 y_0 + 0.5x^T y + e^T(y_0 x + x_0 y) + z_r[-1, 1], \tag{2.22a}$$

$$z_r = |y_0|x_r + |x_0|y_r + (\|x\|_1 + x_r) \cdot (\|y\|_1 + y_r) - 0.5|x|^T|y|. \tag{2.22b}$$

The number of multiplications in (2.22) is $4n + 6$, which is almost twice as much as in the case of trivial bounds, however, this is the cost of smaller overestimation. The number of additions (subtraction is counted as addition) is $4n + 3$, which ultimately gives asymptotic time complexity $\mathcal{O}(n)$, where n is the number of noise symbols. We will refer to multiplication formula (2.22) as *standard multiplication*.

Miyajima [142] proposed a modification on the standard multiplication of affine forms, which gives tighter bounds (however, at even higher cost). The similar modification applied to (2.22) results in:

$$\hat{z} = x_0 y_0 + 0.5x^T y + e^T(y_0 x + x_0 y) + z_r[-1, 1], \tag{2.23a}$$

$$z_r = x_r y_r + x_r(|y_0| + \|y\|_1) + y_r(|x_0| + \|x\|_1) + 0.5|x|^T|y| + \\ + \sum_{i,j=1, i<j}^{n} |x_i y_j + x_j y_i|. \tag{2.23b}$$

The number of multiplications in (2.23) is $n^2 + 3n + 6$, and the number of additions is $n^2 + 4n + 2$. This gives asymptotic time complexity $\mathcal{O}(n^2)$, where n is the number of noise symbols.

Lemma 2.6 *Let $e \in \boldsymbol{U}^n$. Then*

$$\left(e^T x\right)\left(e^T y\right) \in 0.5x^T y + 0.5|x|^T|y| + \sum_{i,j=1, i<j}^{n} |x_i y_j + x_j y_i| \subseteq \\ \subseteq 0.5x^T y + \left(\|x\|_1\|y\|_1 - 0.5|x|^T|y|\right)[-1, 1].$$

Proof We have

$$\left(e^T x\right)\left(e^T y\right) = \sum_{i=1}^{n} x_i y_i \varepsilon_i^2 + \sum_{i,j=1, i<j}^{n} (x_i y_j + x_j y_i)\varepsilon_i \varepsilon_j \in \\ \in \sum_{i=1}^{n} x_i y_i [0, 1] + \sum_{i,j=1, i<j}^{n} |x_i y_j + x_j y_i|[-1, 1] = \\ = 0.5x^T y + \left(0.5|x|^T|y| + \sum_{i,j=1, i<j}^{n} |x_i y_j + x_j y_i|\right)[-1, 1] \subseteq \\ \subseteq 0.5x^T y + \left(\|x\|_1\|y\|_1 - 0.5|x|^T|y|\right)[-1, 1].$$

□

Proposition 2.7 *The affine multiplication defined by (2.22) and (2.23) are inclusion functions of $f(x, y) = x \cdot y$ over joint range $\langle \hat{x}, \hat{y} \rangle$.*

Proof If $\tilde{x} \in [\hat{x}]$ and $\tilde{y} \in [\hat{y}]$, then

$$\tilde{x} = x_0 + e^T x + x_r \varepsilon_x,$$
$$\tilde{y} = y_0 + e^T y + y_r \varepsilon_y,$$

for some $\varepsilon_x, \varepsilon_y \in [-1, 1]$, and, in view of Lemma 2.6, we have

$$\begin{aligned}
\tilde{x} \cdot \tilde{y} &= \left(x_0 + e^T x + x_r \varepsilon_x\right)\left(y_0 + e^T y + y_r \varepsilon_y\right) = \\
&= x_0 y_0 + e^T (x_0 y_i + y_0 x_i) + \left(e^T x\right)\left(e^T y\right) + \\
&+ x_0 y_r \varepsilon_x + \left(e^T x\right) y_r \varepsilon_y + y_0 x_r \varepsilon_x + \left(e^T y\right) x_r \varepsilon_x + x_r y_r \varepsilon_x \varepsilon_y \in \\
&\in x_0 y_0 + e^T (x_0 y_i + y_0 x_i) + 0.5 x^T y + \\
&+ \left(0.5 |x|^T |y| + \sum_{i,j=1, i<j}^{n} |x_i y_j + x_j y_i|\right)[-1, 1] + \\
&+ (|x_0| y_r + \|x\|_1 y_r + |y_0| x_r + \|y\|_1 x_r + x_r y_r)[-1, 1] \subseteq \\
&\subseteq x_0 y_0 + e^T (x_0 y_i + y_0 x_i) + 0.5 x^T y + \left(\|x\|_1 \|y\|_1 - 0.5 |x|^T |y|\right)[-1, 1] + \\
&+ (|x_0| y_r + \|x\|_1 y_r + |y_0| x_r + \|y\|_1 x_r + x_r y_r)[-1, 1].
\end{aligned}$$

After rearranging the terms, we obtain

$$\begin{aligned}
\tilde{x} \cdot \tilde{y} &\in x_0 y_0 + 0.5 x^T y + e^T (x_0 y_i + y_0 x_i) + \\
&+ \left(x_r y_r + x_r (|y_0| + \|y\|_1) + y_r (|x_0| + \|x\|_1)\right. \\
&\left. + 0.5 |x|^T |y| + \sum_{i,j=1, i<j}^{n} |x_i y_j + x_j y_i|\right)[-1, 1] \subseteq \\
&\subseteq x_0 y_0 + 0.5 x^T y + e^T (x_0 y_i + y_0 x_i) + \\
&\left(|y_0| x_r + |x_0| y_r + (\|x\|_1 + x_r) \cdot (\|y\|_1 + y_r) - 0.5 |x|^T |y|\right)[-1, 1],
\end{aligned}$$

which completes the proof. □

Formula (2.23) can be further improved by splitting vector $x^T y$ into positive and negative parts (cf. [235]). It holds that

$$\sum_{i=1}^{n} x_i y_i \varepsilon_i^2 \in \left(\sum_{x_i y_i \geqslant 0} x_i y_i \right) [0, 1] + \left(\sum_{x_i y_i < 0} x_i y_i \right) [0, 1] \subseteq$$
$$\subseteq \max \left\{ \sum_{x_i y_i \geqslant 0} x_i y_i, - \sum_{x_i y_i < 0} x_i y_i \right\} [-1, 1]. \tag{2.24}$$

The resulting revised affine form $\hat{z}$ is given by:

$$\hat{z} = x_0 y_0 + 0.5 x^T y + e^T (y_0 x + x_0 y) + e_z[-1, 1], \tag{2.25a}$$
$$z_r = x_r y_r + x_r (|y_0| + \|y\|_1) + y_r (|x_0| + \|x\|_1) +$$
$$+ 0.5 \max \left\{ \sum_{x_i y_i \geqslant 0} x_i y_i, - \sum_{x_i y_i < 0} x_i y_i \right\} + \sum_{i,j=1, j<i}^{n} |x_i y_j + x_j y_i|. \tag{2.25b}$$

The number of real multiplications and additions is the same as in (2.23).

Proposition 2.8 *The affine multiplications defined by (2.25) is an inclusion function for $f(x, y) = x \cdot y$ over joint range $\langle \hat{x}, \hat{y} \rangle$.*

Proof Follows directly from Proposition 2.7 and (2.24). □

Now, let us write function (2.20) in the following form:

$$f^*(e) = x_0 y_0 + e^T (x_0 y + y_0 x) + x_0 y_r[-1, 1] + y_0 x_r[-1, 1] \tag{2.26}$$
$$+ (e^T x + x_r[-1, 1]) \cdot (e^T y + y_r[-1, 1]),$$

and let $R_{\min}$ and $R_{\max}$ denote the minimum and maximum, respectively, of the last term in (2.26) on its domain. Then, the Chebyshev minimum-error approximation (Fig. 2.5) of the product of the revised affine forms is given by

$$\hat{z} = x_0 y_0 + 0.5(R_{\min} + R_{\max}) + e^T (y_0 x + x_0 y) + z_r[-1, 1], \tag{2.27a}$$
$$z_r = |y_0| x_r + |x_0| y_r + 0.5(R_{\max} - R_{\min}). \tag{2.27b}$$

Approximation (2.27) is computed at the cost of $2n + 5$ multiplications and $n + 3$ additions plus the cost of computing $R_{\min}$ and $R_{\max}$ (see Algorithm 2).

Proposition 2.9 (Skalna and Hladík [258]) *The affine multiplication defined by (2.27) is an inclusion function of $f(x, y) = x \cdot y$ over joint range $\langle \hat{x}, \hat{y} \rangle$.*

Proof Follows directly from the construction of (2.27). □

In order to find $R_{\min}$ and $R_{\max}$, we consider the following function:

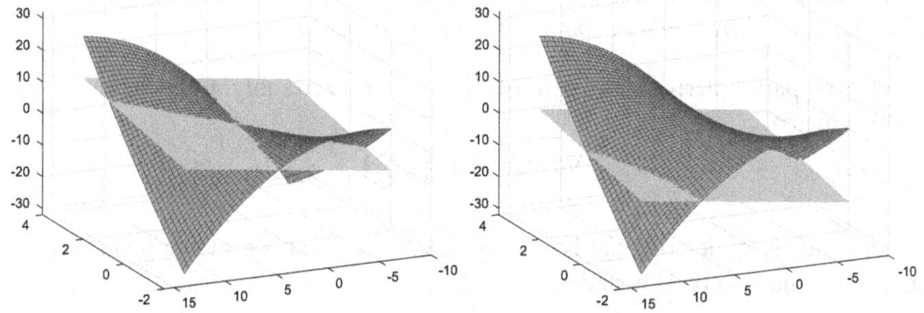

Fig. 2.5 Graph of function $f(x, y) = x \cdot y$ with trivial affine approximation (left) and Chebyshev minimum-error affine approximation (right)

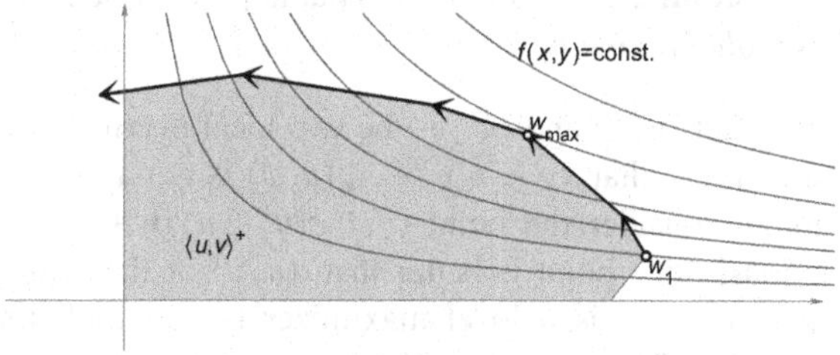

Fig. 2.6 Set $\langle \hat{u}, \hat{v} \rangle^+$ (gray region) with northwest boundary (black arrows), level sets of objective function (2.28) (grey lines), right-most vertex w_1 and point $w_{\max}$ at which maximum $R_{\max}$ is attained

$$R(e, \varepsilon_x, \varepsilon_y) = (e^T x + x_r \varepsilon_x) \cdot (e^T y + y_r \varepsilon_y), \tag{2.28}$$

where $\varepsilon_x, \varepsilon_y$ vary within interval $[-1, 1]$, independently from other noise symbols and from each other. Then,

$$R_{\min} = \min \left\{ R(e, \varepsilon_x, \varepsilon_y) \mid (e, \varepsilon_x, \varepsilon_y) \in \boldsymbol{U}^{n+2} \right\}, \tag{2.29a}$$

$$R_{\max} = \max \left\{ R(e, \varepsilon_x, \varepsilon_y) \mid (e, \varepsilon_x, \varepsilon_y) \in \boldsymbol{U}^{n+2} \right\}. \tag{2.29b}$$

By using the joint range notation (see Sect. 2.1), problem (2.29b) can be formulated as the following constrained optimization problem:

$$\begin{aligned} \max \quad & u \cdot v \\ \text{subject to} \quad & (u, v) \in \left\langle e^T x + x_r \varepsilon_x, e^T y + y_r \varepsilon_y \right\rangle. \end{aligned} \tag{2.30}$$

Since both the function to be maximized and the joint range $\langle \hat{u}, \hat{v} \rangle$ are symmetric about the origin, the maximum must be attained at some point lying in non-negative quadrant (Fig. 2.6)

$$\langle \hat{u}, \hat{v} \rangle^+ \triangleq \{(u, v) \in \left\langle \hat{u}, \hat{v} \right\rangle \mid u, v \geqslant 0\}.$$

Moreover, as is shown in the next proposition, it is enough to search boundary $\partial \langle \hat{u}, \hat{v} \rangle$ of joint range $\langle \hat{u}, \hat{v} \rangle$.

Proposition 2.10 *Let $\hat{u} = e^T x + x_r \varepsilon_x$, $\hat{v} = e^T y + y_r \varepsilon_y$. Then, function $f(u, v) = u \cdot v$ attains its extrema on the boundary of joint range $\langle \hat{u}, \hat{v} \rangle$.*

Proof First partial derivatives $\frac{\partial f}{\partial u}(u, v)$, $\frac{\partial f}{\partial v}(u, v)$ are zero only at $(0, 0)$. However, Hessian matrix

$$\nabla^2 f(0, 0) = \begin{pmatrix} 0 & 1 \\ 1 & 0 \end{pmatrix}$$

is indefinite, as its determinant is -1, so the critical point is a saddle point, hence, the extrema have to be at boundary. □

Further, each local maximizer lying in $\langle \hat{u}, \hat{v} \rangle^+$ is a global optimum. Thus, its is sufficient to seek for a local maximizer only [258].

Proposition 2.11 (Skalna and Hladík [258]) *Each local maximizer lying in $\langle \hat{u}, \hat{v} \rangle^+$ is a global optimum.*

Proof Let (u, v) and (u', v') be two local maximizers lying in $\langle \hat{u}, \hat{v} \rangle^+$. Suppose to the contrary that $uv < u'v'$, i.e., (u, v) is not a global maximizer. For $\varepsilon > 0$ small enough consider the point $(u + \varepsilon(u' - u), v + \varepsilon(v' - v))$. If $u \leqslant u'$ and $v \leqslant v'$, obviously this point is better than (u, v) or the same for every $\varepsilon > 0$, contradicting that (u, v) is a local maximizer (or contradicting $uv < u'v'$ in the case of equality). Thus assume without loss of generality that $u \leqslant u'$ and $v \geqslant v'$. This implies $(u - u')(v' - v) \geqslant 0$, whence $uv + u'v' \leqslant uv' + u'v$. Now, the objective value in that point is $(u + \varepsilon(u' - u))(v + \varepsilon(v' - v)) = uv + \varepsilon^2(u' - u)(v' - v) + \varepsilon(uv' + u'v - 2uv) \geq uv + \varepsilon^2(u' - u)(v' - v) + \varepsilon(u'v' - uv) > uv$ for any $\varepsilon > 0$ small enough. This contradicts that (u, v) is a local maximizer. □

In [258], we have proposed an efficient method (see Algorithm 2) for computing extremal values $R_{\min}$, $R_{\max}$. The general idea of this algorithm is the following: start at right-most vertex w_1 of $\langle \hat{u}, \hat{v} \rangle$ and follow the boundary of $\langle \hat{u}, \hat{v} \rangle^+$ in the northwest direction until the maximizer is found. To follow the boundary of $\langle \hat{u}, \hat{v} \rangle^+$, we compute slopes

$$s_i := \frac{\operatorname{sgn}(x_i) y_i}{|x_i|}, \quad i = 1, \ldots, n+2, \ x_i \neq 0, \tag{2.31}$$

where

$$\operatorname{sgn}(x) = \begin{cases} 1, & x \geqslant 0, \\ -1, & x < 0. \end{cases} \tag{2.32}$$

Next, the median of the sequence (2.31) is found. Based on the objective values of the median and its neighbors, one half of the sequence can be removed. Repeating this procedure, the optimum is attained in a linear time, which is optimal.

Thus, the overall asymptotic time complexity of computing (2.27) is $\mathcal{O}(n)$, where n is the number of noise symbols.

Remark The $\mathcal{O}(n^2)$ algorithm for computing the extremal values of function $f(x, y) = x \cdot y$ on joint range $\langle \hat{x}, \hat{y} \rangle$ was proposed in [144], however, the algorithm does not always yield the correct result. In [254], we have proposed a correction that prevents the method from producing incorrect results.

Algorithm 2 (cf. **[258]**) Computing maximum $R_{\max}$ of function (2.28)

Input: $\hat{x} = \sum_{i=1}^{n} x_i \varepsilon_i + x_r \varepsilon_x$, $\hat{y} = \sum_{i=1}^{n} y_i \varepsilon_i + y_r \varepsilon_y$,
Output: $R_{\max} = \max\{R(e, \varepsilon_x, \varepsilon_y), (e, \varepsilon_x, \varepsilon_y) \in \boldsymbol{U}^{n+2}\}$
$x := (x_1, \ldots, x_n, x_r, 0)$, $y := (y_1, \ldots, y_n, 0, y_r)$
$y_i := |y_i|$ whenever $x_i = 0, i = 1, \ldots, n+2$
$w_1 := (\|x\|_1, \mathrm{sgn}(x)^T y)$ // the right-most vertex
$s_i := \mathrm{sgn}(x_i) y_i / |x_i|, i = 1, \ldots, n+2, x_i \neq 0$
$h_i := -2(|x_i|, \mathrm{sgn}(x_i) y_i), i = 1, \ldots, n+2, x_i \neq 0$
let $i_k, i_{k'}, i_{k''}$ denote the index of the median and its left and right neighboring values, respectively
$(u_k, v_k) := w_1 + \sum_{i: s_i \leqslant s_{i_k}} h_i$
$(u_{k'}, v_{k'}) := w_1 + \sum_{i: s_i \leqslant s_{i_{k'}}} h_i$
$(u_{k''}, v_{k''}) := w_1 + \sum_{i: s_i \leqslant s_{i_{k''}}} h_i$
if $(u_k \cdot v_k < u_{k'} \cdot v_{k'})$ **then**
recursively resolve the problem on subset $\{s_i \mid s_i \leqslant s_{i_k}\}$
else if $(u_k \cdot v_k < u_{k''} \cdot v_{k''})$ **then**
$w_1 := w_1 + (u_k, v_k)$
recursively resolve the problem on subset $\{s_i \mid s_i \geqslant s_{i_k}\}$
else
inspect both edges emerging from point (u_k, v_k) for a possible maximizer
end if
return $R_{\max}$

Proposition 2.12 *The following inclusions hold:*

$$[\hat{z}]_{(2.27)} \subseteq [\hat{z}]_{(2.25)} \subseteq [\hat{z}]_{(2.23)} \subseteq [\hat{z}]_{(2.22)} \subseteq [\hat{z}]_{(2.21)}. \tag{2.33}$$

Proof We have

$$0.5 \max \left\{ \sum_{x_i y_i \geqslant 0} x_i y_i, -\sum_{x_i y_i < 0} x_i y_i \right\} \leqslant 0.5|x|^T|y|.$$

Hence

$$0.5x^T y + 0.5 \max \left\{ \sum_{x_i y_i \geqslant 0} x_i y_i, -\sum_{x_i y_i < 0} x_i y_i \right\} + \sum_{i,j=1, i<j}^{n} |x_i y_j + x_j y_i| \leqslant$$

$$\leqslant 0.5x^T y + 0.5|x|^T|y| + \sum_{i,j=1, i<j}^{n} |x_i y_j + x_j y_i| \leqslant$$

$$\leqslant 0.5x^T y + 0.5|x|^T|y| + \sum_{i,j=1,i\neq j}^{n} |x_i y_j| =$$
$$= 0.5x^T y + \|x\|_1\|y\|_1 - 0.5|x|^T|y| \leqslant \|x\|_1\|y\|_1.$$

So, $\sup([\hat{z}]_{(2.25)}) \leqslant \sup([\hat{z}]_{(2.23)}) \leqslant \sup([\hat{z}]_{(2.22)}) \leqslant \sup([\hat{z}]_{(2.21)})$. The respective inequalities for the lower bounds can be shown in a similar manner. The first inclusion in (2.33) follows from the fact that $[\hat{z}]_{(2.27)}$ is the optimal revised affine form. □

According to de Figueiredo and Stolfi [33], the optimal multiplication form can be tighter when compared to trivial form (2.21) up to a factor of 4 (this occurs, e.g., for $\hat{x} = \varepsilon_1 + \varepsilon_2$, $\hat{y} = \varepsilon_1 - \varepsilon_2$). Moreover, despite the fact that form (2.22) is at least as good as (2.21), it is also four-times wider than the optimal form in the worst case. This is illustrated by the following example:

Example 2.13 Let $x = (1, \ldots, 1)^T \in \mathbb{R}^{2n}$ and $y = (1, -1, \ldots, 1, -1)^T \in \mathbb{R}^{2n}$. Then, the enclosure of quadratic form $(e^T x)(e^T y)$, $e \in \boldsymbol{U}^{2n}$, computed by (2.22), has a width of $2n^2 - n$. To determine the optimal width, the following substitution is used:

$$x'_1 := \frac{2}{n}(x_1 + x_3 + \cdots + x_{2n-1}) = 1,$$
$$x'_2 := \frac{2}{n}(x_2 + x_4 + \cdots + x_{2n}) = 1,$$
$$y'_1 := \frac{2}{n}(y_1 + y_3 + \cdots + y_{2n-1}) = 1,$$
$$y'_2 := \frac{2}{n}(y_2 + y_4 + \cdots + y_{2n}) = -1,$$

and the quadratic form is rewritten as

$$(e^T x)(e^T y) = \frac{n^2}{4}(e'^T x')(e'^T y'),$$

where $e' \in \boldsymbol{U}^2$. It is easy to determine analytically that quadratic form $(e'^T x')(e'^T y')$ ranges in $[-1, 1]$, so original quadratic form $(e^T x)(e^T y)$ ranges in $[-\frac{n^2}{4}, \frac{n^2}{4}]$, having a width of $\frac{n^2}{2}$. Therefore, the overestimation factor is

$$\frac{2n^2 - n}{n^2/2} = 4 - \frac{2}{n},$$

which converges to 4 as $n \to \infty$.

The following two examples compare the performance of the presented multiplication formulae.

Table 2.1 Results for Example 2.14: number of iterations (#Itrs) and number of boxes left in list L (#Bxs)

Function	(2.21)		(2.22), (2.23), (2.25)		(2.27)	
	#Itrs	#Bxs	#Itrs	#Bxs	#Itrs	#Bxs
Colville	266034	17543	171484	14425	95871	10383
Dixon and Price ($n = 5$)	242190	34950	121142	25091	72755	18019
Rosenbrock ($n = 4$)	167560	14451	104927	11010	62601	8442
Average	225261	22315	132518	16842	77076	12281

Example 2.14 In this example the Moore–Skelboe algorithm (cf. [197]) is used to optimize three benchmark function (see, e.g., [285–287]). The range of test functions is computed using revised affine arithmetic.

- *Colville function*

$$\begin{aligned} f_1(x) &= 100(x_1^2 - x_2)^2 + (x_1 - 1)^2 + (x_3 - 1)^2 + 90(x_3^2 - x_4)^2 \\ &\quad + 10.1((x_2 - 1)^2 + (x_4 - 1)^2) + 19.8(x_2 - 1)(x_4 - 1),\ x \in [-10, 10]^4 \end{aligned}$$

- *Dixon* and *Price* (scalable)

$$f_2(x) = (x_1 - 1)^2 + \sum_{i=2}^{n} i(2x_i^2 - x_{i-1})^2,\ x \in [-n, n]^n$$

- *Rosenbrock* (scalable)

$$f_3(x) = \sum_{i=1}^{n-1} (100(x_{i+1} - x_i^2)^2 + (x_i - 1)^2),], x \in [-10, 10]^n$$

Table 2.1 shows that multiplication (2.27) always gives the smallest number of iterations (#Itrs) and the smallest number of elements remaining in the list (#Bxs) at the end of the computation. Comparing with the other methods, an impressive gain of about 40% can be noticed. The CPU times are listed in Table 2.2. As can be seen, the average gain in CPU time is not less than 26%.

Example 2.15 This example compares the performance and efficiency of different multiplication methods for drawing implicit polynomial curves on a rectangular grid of pixels. The implicit polynomial curve for $f(x, y)$ is defined as set of points

$$\{(x, y) \mid f(x, y) = 0\}.$$

Table 2.2 Results for Example 2.14: CPU times (in seconds) and ratio of (2.27) time to the best time

Function	(2.21)	(2.22)	(2.23)	(2.25)	(2.27)	(2.27)/t_{best}
Colville	294.39	164.03	156.01	156.33	99.82	1.0
Dixon and Price ($n = 5$)	431.96	152.41	142.71	147.54	121.53	1.0
Rosebrock ($n = 4$)	164.79	100.26	96.85	98.40	70.07	1.0
Average	297.05	138.90	131.86	134.09	97.14	1.0

Drawing such curves is of great interest in CAD, CAGD and computer graphics [135]. The following two polynomials functions are considered:

- Symmetric medium degree polynomial (Ratschek and Rokne) [135]:

$$\begin{aligned}&47.6 - 220.8x + 476.8x^2 - 512x^3 + 256x^4 - 220.8y + 512xy\\&- 512x^2y + 476.8y^2 - 512xy^2 + 512x^2y^2 - 512y^3 + 256y^4\end{aligned} \tag{2.34}$$

- Asymmetric medium degree polynomial [135]:

$$\begin{aligned}&(55/256) - x + 2x^2 - 2x^3 + x^4 - (55/64)y + 2xy - 2x^2y+\\&(119/64)y^2 - 2xy^2 + 2x^2y^2 - 2y^3 + y^4\end{aligned} \tag{2.35}$$

The recursive procedure *Quadtree* (cf. [135]) is used to plot the curves on a grid of 256×256 pixels. The basic strategy of the procedure is to evaluate function $f(x, y)$ on a box. If the resulting interval does not contain zero, the box is discarded, otherwise it is subdivided into four smaller boxes and the process is repeated for each box. The process stops and the pixel is filled when the box contains only one pixel. The plots of curves (2.34) and (2.35) for $(x, y) \in [0, 1] \times [0, 1]$, obtained using the above described procedure, are shown in Figs. 2.7 and 2.8.

To compare the performance and efficiency of the different multiplication methods for curve plotting, the following quantities were measured: the number of pixels filled, the number of subdivisions involved, the number of affine additions and multiplications performed in evaluating the functions. As justified in [135], the smaller these values are, the better. The results are presented in Tables 2.3 and 2.4 together with the computational times. As can be seen, the proposed here Chebyshev multiplication (2.27) always yields the smallest number of pixels, subdivisions, additions and multiplications, but has the worst computational time. From among the remaining methods, the standard multiplication (2.22) is the best as it gives the same number of pixels, subdivisions, additions and multiplications as (2.23) and (2.25), but is a bit faster. Whereas the simple multiplication (2.21) is the slowest and it gives the

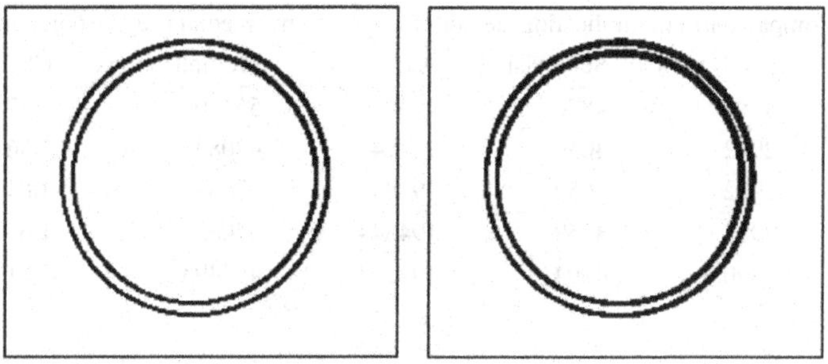

Fig. 2.7 Symmetric medium degree polynomial: $47.6 - 220.8x + 476.8x^2 - 512.0x^3 + 256.0x^4 - 220.8y + 512.0xy - 512.0x^2y + 476.8y^2 - 512.0xy^2 + 512.0x^2y^2 - 512.0y^3 + 256.0y^4$: plot obtained using multiplication (2.27) (left); plot obtained using trivial multiplication (2.21) (right)

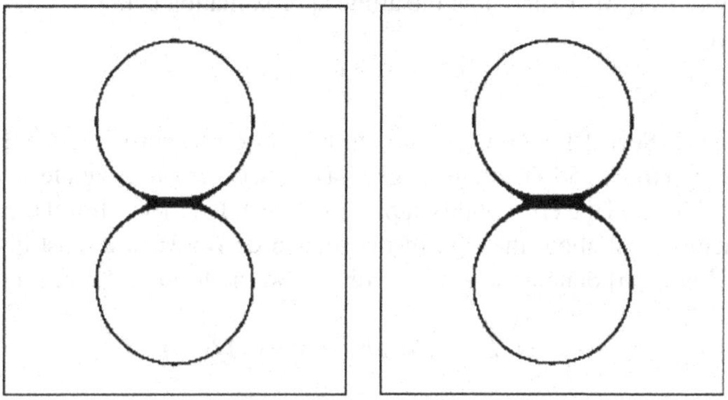

Fig. 2.8 Asymmetric medium degree polynomial: $((55.0/256.0) - x + 2.0x^2 - 2.0x^3 + x^4 - (55.0/64.0)y + 2.0xy - 2.0x^2y + (119.0/64.0)y^2 - 2.0xy^2 + 2.0x^2y^2 - 2.0y^3 + y^4)$: plot obtained using the multiplication (2.27) (left); plot obtained using trivial multiplication (2.21) (right)

Table 2.3 Comparison of multiplication methods for symmetric medium degree polynomial (2.34)

Method	Pixels plotted	Subdivisions	Additions	Multiplications	CPU time [s]
(2.21)	8651	9584	460044	1226784	3.84
(2.22)	6262	8387	402156	1072416	3.35
(2.23)	6262	8387	402156	1072416	3.51
(2.25)	6262	8387	402156	1072416	3.52
(2.27)	6116	8307	398748	1063328	5.28

largest values of the selected characteristics, which is reflected in lower quality of the respective drawings (Fig. 2.7(right) and Fig. 2.8(right)).

Table 2.4 Comparison of multiplication methods for asymmetric medium degree polynomial (2.35)

Method	Pixels plotted	Subdivisions	Additions	Multiplications	CPU time [s]
(2.21)	3386	4876	233628	564601	1.87
(2.22)	2922	4059	194844	470837	1.56
(2.23)	2922	4059	194844	470837	1.64
(2.25)	2922	4059	194844	470837	1.65
(2.27)	2890	4004	192204	464493	2.51

2.2.2 Reciprocal

Since reciprocal function $f(x) = 1/x$ is defined for $x \neq 0$, we can assume that $0 \notin [\hat{x}]$. Moreover, we assume that the affine approximation being searched has the following form:

$$f^a(x) = \alpha x + \gamma. \tag{2.36}$$

As mentioned in Sect. 2.1.2, the Chebyshev minimum-error approximation (Fig. 2.9a) has the required form, and Theorem 2.2 gives a recipe on how to compute coefficients α and γ in (2.36) and the error of this approximation δ. In order to fulfill the assumption of Theorem 2.2 about the sign of the second derivative, we must distinguish two cases. Let $[a, b]$ denote the range of revised affine form $\hat{x}$. Then, according to (2.11a),

$$\alpha = (1/b - 1/a)/(b - a) = -1/(ab). \tag{2.37}$$

So, $f'(u) = \alpha$ is satisfied with $u = \pm\sqrt{ab}$ and from (2.12) it follows that $r(u) = (a + b - \sqrt{ab})/(ab)$. Finally, (2.11b) and (2.11c) yield:

$$\gamma = \begin{cases} \dfrac{a + b + 2\sqrt{ab}}{2ab}, & a > 0, \\ \dfrac{a + b - 2\sqrt{ab}}{2ab}, & b < 0, \end{cases} \tag{2.38a}$$

$$\delta = \begin{cases} \dfrac{a + b - 2\sqrt{ab}}{2ab}, & a > 0, \\ \dfrac{a + b + 2\sqrt{ab}}{2ab}, & b < 0, \end{cases} \tag{2.38b}$$

and revised affine form $\hat{z}$ that describes $1/\hat{x}$ is given by

$$\hat{z} = \alpha x_0 + \gamma + e^T(\alpha x) + (|\alpha| x_r + \delta)[-1, 1]. \tag{2.39}$$

The computation of the Chebyshev minimum-error approximation of reciprocal $1/\hat{x}$ requires $n + 8$ multiplications, 4 additions, and a square root to be computed, which means that the asymptotic time complexity is $\mathcal{O}(n)$, where n is the number of noise symbols.

Proposition 2.16 *Affine reciprocal $\hat{z}$ defined by (2.38), and (2.39) is an inclusion function of $f(x) = 1/x$.*

Proof Since $0 \notin [a, b]$, revised affine form (2.38), (2.39) is well-defined. If $x \in \hat{x}$, then $\varepsilon_x \in [-1, 1]$ exists such that $x = x_0 + e^T x + x_r \varepsilon_x$, and from the construction of α, γ, δ, it follows that

$$\begin{aligned}1/x &\in \alpha x + \gamma + \delta[-1, 1] = \alpha(x_0 + e^T x + x_r \varepsilon_x) + \delta[-1, 1] = \\ &= \alpha x_0 + e^T(\alpha x) + \alpha x_r \varepsilon_x + \delta[-1, 1] \in \\ &\in \alpha x_0 + e^T(\alpha x) + (|\alpha| x_r + \delta)[-1, 1] \subseteq [\hat{z}].\end{aligned}$$

□

According to Corollary 2.3, the range of revised affine form $\hat{z}$ overestimates the lower ($[\hat{x}] > 0$) or the upper ($[\hat{x}] < 0$) bound of the true range of $1/\hat{x}$ by $(\sqrt{a} - \sqrt{b})^2/ab$ ($(\sqrt{a} + \sqrt{b})^2/ab$). However, it follows from our computational experience in solving parametric interval linear systems that, despite the overestimation, the use of the Chebyshev minimum-error approximation of the reciprocal to define the division of the revised affine forms guarantees tighter bounds for the parametric solution set (see Chap. 4) than the use of the min-range approximation.

Nevertheless, in general, it is recommended that the reciprocal function use the min-range approximation instead of the Chebyshev one (see [33]). In order to obtain the min-range approximation (Fig. 2.9b), coefficients α and γ must be chosen so that joint range $\langle \hat{x}, \alpha\hat{x} + \gamma + \delta\varepsilon_z \rangle$ (where $\delta\varepsilon_z$ represents the approximation error) has the same vertical extent as the piece of the graph of the reciprocal function (cf. [33]). Since the reciprocal function is convex monotone decreasing in each interval $[a, b] > 0$ and concave monotone decreasing in each interval $[a, b] < 0$, the "min-range" revised affine form that describes $1/\hat{x}$ over interval $[a, b]$ is defined by (cf. Proposition 2.4)

$$\alpha = \begin{cases} -\dfrac{1}{b^2}, & a > 0, \\ -\dfrac{1}{a^2}, & b < 0, \end{cases} \tag{2.40a}$$

$$\gamma = \begin{cases} \dfrac{(a+b)^2}{2ab^2}, & a > 0, \\ \dfrac{(a+b)^2}{2a^2 b}, & b < 0, \end{cases} \tag{2.40b}$$

$$\delta = \begin{cases} \dfrac{(b-a)^2}{2ab^2}, & a > 0, \\ \dfrac{(b-a)^2}{2a^2 b}, & b < 0, \end{cases} \tag{2.40c}$$

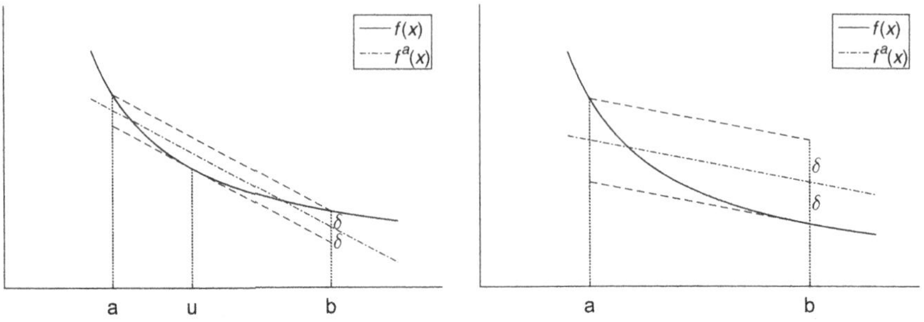

Fig. 2.9 Geometrical illustration of Chebyshev minimum-error (left) and min-range (right) approximations of reciprocal function

The computation of the min-range approximation to the reciprocal function requires $n + 14$ multiplications and 3 additions, which gives us asymptotic time complexity $\mathcal{O}(n)$, where n is the number of noise symbols.

Proposition 2.17 *The affine reciprocal defined by (2.40), (2.39) is the min-range inclusion function of* $f(x) = 1/x$.

Proof Since $0 \notin [a, b]$, revised affine form (2.40), (2.39) is well-defined. The inclusion property of $\hat{z}$ can be proven in a similar manner as in the proof of Proposition 2.16. So, it is enough to show that the range of $\hat{z}$ is minimal. Assume that $a > 0$, then, the true range of $1/x$ over $[a, b]$ is $[1/b, 1/a]$, whereas from (2.38) and (2.39), it follows that

$$\begin{aligned}\inf([\hat{z}]) &= \alpha x_0 - |\alpha| \|x\|_1 - |\alpha| x_r + \gamma - \delta = \\ &= (-1/b^2)(x_0 + \|x\|_1 + x_r) + \gamma - \delta = \\ &= -1/b + 2/b = 1/b,\end{aligned}$$

$$\begin{aligned}\sup([\hat{z}]) &= \alpha x_0 + |\alpha| \|x\|_1 + |\alpha| x_r + \gamma + \delta = \\ &= (-1/b^2)(x_0 - \|x\|_1 - x_r) + \gamma + \delta = \\ &= -a/b^2 + (a^2 + b^2)/(ab^2) = 1/a.\end{aligned}$$

The optimality for $b < 0$ can be proven analogously. □

Any other nonlinear univariate function that satisfies the conditions of Theorem 2.2 and/or Proposition 2.4 can be approximated in a similar manner.

2.2.3 Division

Division, like multiplication, can be done in many ways, however, it is convenient to look at division as a multiplication of the dividend times the reciprocal of the divisor. So, given two revised affine forms $\hat{x}$, $\hat{y}$ ($0 \notin [\hat{y}]$), their quotient $\hat{x}/\hat{y}$ is rewritten as product $\hat{x} * (1/\hat{y})$. This approach has quadratic convergence, since the reciprocal procedure preserves first order correlations between $\hat{x}$ and $\hat{y}$ [33]. The division of affine forms that employs Chebyshev minimum-error multiplication was described in [143].

However, the thus-defined division does not guarantee that quotient $\hat{x}/\hat{x}$ will result in an affine form having a range close to 1. In order to obtain such a property, the revised affine form $\hat{z}$ that represents the quotient of revised affine forms $\hat{x}$, $\hat{y}$ can be defined as follows (cf. [45, 72, 106]):

$$\hat{z} = x_0/y_0 + \hat{p} * (1/\hat{y}), \tag{2.41a}$$

$$\hat{p} = e^T(x - (x_0/y_0)y) + (x_r + |x_0/y_0|y_r)[-1, 1]. \tag{2.41b}$$

In order to compute revised affine form $\hat{p}$, we must perform $n + 1$ multiplications and $n + 1$ additions. So, the asymptotic time complexity of division (2.41) depends on the multiplication method used in (2.41a). If $x_r = 0$, then $\hat{x}/\hat{x} = 1$ (assuming exact computation).

Proposition 2.18 *The division of revised affine forms $\hat{x}$ and $\hat{y}$, $0 \notin [\hat{y}]$ defined by (2.41) is an inclusion function for the real division.*

Proof Since $0 \notin [\hat{y}]$, formulae (2.41a) and (2.41b) are well-defined. Now, it is enough to show that, for each $x \in [\hat{x}]$, $y \in [\hat{y}]$, it holds that $x/y \in [\hat{z}]$. By definition, $\varepsilon_x \in [-1, 1]$, $\varepsilon_y \in [-1, 1]$ and e exist such that $x = x_0 + e^T x + x_r\varepsilon_x$ and $y = y_0 + e^T y + y_r\varepsilon_y$. Therefore, we have

$$\begin{aligned}
\frac{x}{y} &= \frac{x_0 + e^T x + x_r\varepsilon_x}{y_0 + e^T y + y_r\varepsilon_y} = \frac{y_0(x_0 + e^T x + x_r\varepsilon_x)}{y_0(y_0 + e^T y + y_r\varepsilon_y)} = \\
&= \frac{x_0(y_0 + e^T y + y_r\varepsilon_y) + e^T(y_0 x - x_0 y) + y_0 x_r\varepsilon_x - x_0 y_r\varepsilon_y}{y_0(y_0 + e^T y + y_r\varepsilon_y)} = \\
&= \frac{x_0}{y_0} + \frac{e^T(y_0 x - x_0 y) + y_0 x_r\varepsilon_x - x_0 y_r\varepsilon_y}{y_0(y_0 + e^T y + y_r\varepsilon_y)} = \\
&= \frac{x_0}{y_0} + \frac{e^T(x - (x_0/y_0)y) + x_r\varepsilon_x - (x_0/y_0)y_r\varepsilon_y}{y_0 + e^T y + y_r\varepsilon_y} \in \\
&\in \frac{x_0}{y_0} + \frac{e^T(x - (x_0/y_0)y) + (x_r + |x_0/y_0|y_r)[-1, 1]}{y_0 + e^T y + y_r[-1, 1]} \subseteq [\hat{z}].
\end{aligned}$$

□

2.2.4 Rounded Revised Affine Arithmetic

In order to obtain rigorous results using affine arithmetic, each operation on revised affine forms must take into account rounding errors. However, it is not enough to round each coefficient z_i of $\hat{z}$ in the "safe" direction, since in AA there is no "safe" direction [33]. If noise variable ε_i occurs in some other affine form $\hat{x}$, then any change in z_i would imply different correlation between the quantities represented by $\hat{z}$ and $\hat{x}$, and what follows would falsify the fundamental invariant of AA [33].

In order to preserve the fundamental invariant of AA, basic arithmetic operations on revised affine forms can be performed in the following way. Given $\hat{x}$ and $\hat{y}$, coefficients z_i, u_i and d_i of resulting revised affine form $\hat{z}$ are computed in rounding to nearest, round up and round down modes, respectively. Then, the maximum total rounding error is added to accumulative error z_r in round up mode. For example, the addition of two revised affine forms can look like this:

Algorithm 3 (Addition of revised affine forms)

RAA operator+(RAA $\hat{x}$, RAA $\hat{y}$)
$z_0 = x_0 + y_0$
// Ind($\hat{x}$) – set of indices of $\hat{x}$
for ($i \in$Ind($\hat{x}$)$\cup$Ind($\hat{y}$)) **do**
  $z_i = x_i + y_i$
end for
SetRoundDownMode
$d_0 = x_0 + y_0$
for ($i \in$Ind($\hat{x}$)$\cup$Ind($\hat{y}$)) **do**
  $d_i = x_i + y_i$
end for
SetRoundUpMode
$u_0 = x_0 + y_0$
$\delta = \max\{u_0 - z_0, z_0 - d_0\}$
for ($i \in$Ind($\hat{x}$)$\cup$Ind($\hat{y}$)) **do**
  $u_i = x_i + y_i$
  $\delta = \delta + \max\{u_i - z_i, z_i - d_i\}$
end for
return $\hat{z} = z_0 + e^T z + (\delta + x_r + y_r)[-1, 1]$

This approach is simple but rather expensive. Possible modifications are described in [33]. For example, two multiplications per coefficient can be saved, at the cost of lower precision, by setting $z_i = d_i$ and adding to the accumulative error the difference between u_i and d_i.

Chapter 3
Interval and Parametric Interval Matrices

We start this chapter from an overview of the preliminary theory on interval matrices (cf. [160]), since it lays a basis for our main theoretical considerations. The remaining part of the chapter is devoted to parametric interval vectors and matrices.

3.1 Interval Matrix

An $m \times n$ *interval matrix* is a rectangular array $\boldsymbol{A} = (\boldsymbol{A}_{ij})$ of intervals $\boldsymbol{A}_{ij} \in \mathbb{IR}$, $i = 1, \ldots, m, j = 1, \ldots, n$. If $m = n$, $\boldsymbol{A}$ is said to be square of degree n. The set of all $m \times n$ interval matrices is denoted by $\mathbb{IR}^{m \times n}$. An n-dimensional interval vector $\boldsymbol{b}$ is an n-tuple of intervals $\boldsymbol{b} = (\boldsymbol{b}_i)$, $\boldsymbol{b}_i \in \mathbb{IR}$, $i = 1, \ldots, n$, or simply a one column matrix. The set of all interval vectors is denoted by $\mathbb{IR}^n$. For two interval matrices $\boldsymbol{A}$, $\boldsymbol{B}$ of the same size, the comparison and inclusion relationships are understood component-wise. Similar to the scalar case, certain real matrices are related to each interval matrix $\boldsymbol{A} \in \mathbb{IR}^{m \times n}$. These are:

$$
\begin{aligned}
\inf(\boldsymbol{A}) &= \underline{A} = (\underline{A}_{ij}) && \text{left-endpoint matrix,}\\
\sup(\boldsymbol{A}) &= \overline{A} = (\overline{A}_{ij}) && \text{right-endpoint matrix,}\\
\operatorname{mid}(\boldsymbol{A}) &= A^c = (A^c_{ij}) && \text{midpoint matrix,}\\
\operatorname{rad}(\boldsymbol{A}) &= A^\Delta = (A^\Delta_{ij}) && \text{radius matrix,}\\
|\boldsymbol{A}| &= (|\boldsymbol{A}_{ij}|) && \text{magnitude matrix.}
\end{aligned}
$$

Additionally, with each square interval matrix $\boldsymbol{A}$, we associate a real matrix, the so-called *comparison matrix* [169], with non-negative diagonal elements and non-positive elements outside the diagonal:

$$
\langle \boldsymbol{A} \rangle = \begin{cases} -|\boldsymbol{A}_{ij}|, & i \neq j, \\ \langle \boldsymbol{A}_{ij} \rangle, & i = j. \end{cases} \tag{3.1}
$$

I. Skalna, *Parametric Interval Algebraic Systems*, Studies in Computational Intelligence 766, https://doi.org/10.1007/978-3-319-75187-0_3

The comparison matrix has many important applications in the analysis of both interval linear systems and parametric interval linear systems.

Elementary operations on interval matrices (i.e., addition, subtraction, multiplication, and multiplication by an interval) are defined as follows: [160]. Let $\boldsymbol{A}, \boldsymbol{B} \in \mathbb{IR}^{m\times n}$, $\boldsymbol{C} \in \mathbb{IR}^{n\times p}$ and $\boldsymbol{a} \in \mathbb{IR}$. Then,

$$\begin{aligned}
(\boldsymbol{A}+\boldsymbol{B})_{ij} &= \boldsymbol{A}_{ij}+\boldsymbol{B}_{ij},\\
(\boldsymbol{A}-\boldsymbol{B})_{ij} &= \boldsymbol{A}_{ij}-\boldsymbol{B}_{ij},\\
(\boldsymbol{A}\boldsymbol{C})_{ij} &= \sum_{k=1}^{n}\boldsymbol{A}_{ik}\boldsymbol{C}_{kj},\\
(\boldsymbol{a}\boldsymbol{A})_{ij} &= \boldsymbol{a}\boldsymbol{A}_{ij}.
\end{aligned}$$

The multiplication of interval matrices has the following properties. If $\boldsymbol{A}, \boldsymbol{A}' \in \mathbb{IR}^{m\times n}$, $\boldsymbol{B}, \boldsymbol{B}' \in \mathbb{IR}^{n\times p}$, then

$$\boldsymbol{A}(\boldsymbol{B}+\boldsymbol{B}') \subseteq \boldsymbol{A}\boldsymbol{B}'+\boldsymbol{A}\boldsymbol{B}' \tag{3.2}$$

with equality if $A^{\Delta}=0$ or if $\boldsymbol{B}, \boldsymbol{B}' \geqslant 0$ or if $\boldsymbol{B}, \boldsymbol{B}' \leqslant 0$,

$$\boldsymbol{A}(\boldsymbol{B}\pm\boldsymbol{B}') \subseteq \boldsymbol{A}\boldsymbol{B}\pm\boldsymbol{A}\boldsymbol{B}' \tag{3.3}$$

with equality if $A^{\Delta}=0$,

$$(\boldsymbol{A}\pm\boldsymbol{A}')\boldsymbol{B} \subseteq \boldsymbol{A}\boldsymbol{B}\pm\boldsymbol{A}'\boldsymbol{B} \tag{3.4}$$

with equality if $B^{\Delta}=0$. Additionally

$$\alpha\boldsymbol{A} = \{\alpha A \mid A \in \boldsymbol{A}\} \text{ for all } \alpha \in \mathbb{R}, \tag{3.5}$$

$$\boldsymbol{A}x = \{Ax \mid A \in \boldsymbol{A}\} \text{ for all } x \in \mathbb{R}^n, \tag{3.6}$$

but usually

$$\boldsymbol{A}\boldsymbol{x} \neq \{Ax \mid A \in \boldsymbol{A}, x \in \boldsymbol{x}\}.$$

Selected important properties of interval matrices are summarized below.

Proposition 3.1 (Neumaier [160]) *Let $\boldsymbol{A}, \boldsymbol{B} \in \mathbb{IR}^{n\times n}$. Then,*

$$\boldsymbol{A}' \subseteq \boldsymbol{A} \Rightarrow \langle \boldsymbol{A}'\rangle \geqslant \langle \boldsymbol{A}\rangle, \tag{3.7}$$

$$\langle \boldsymbol{A}\rangle \geqslant \langle A^c\rangle - A^{\Delta}, \text{ with equality iff } 0 \notin \boldsymbol{A}_{ii}, \tag{3.8}$$

$$\langle \boldsymbol{A}\rangle - |\boldsymbol{B}| \leqslant \langle \boldsymbol{A}\pm\boldsymbol{B}\rangle \leqslant \langle \boldsymbol{A}\rangle + |\boldsymbol{B}|, \tag{3.9}$$

$$\langle \boldsymbol{A}\rangle - |\boldsymbol{B}| \leqslant |\boldsymbol{A}\pm\boldsymbol{B}|, \tag{3.10}$$

$$|\boldsymbol{A}\boldsymbol{B}| \geqslant \langle \boldsymbol{A}\rangle|\boldsymbol{B}|, \tag{3.11}$$

$$\langle \boldsymbol{A}\rangle = \langle A\rangle \text{ for some } A \in \boldsymbol{A}. \tag{3.12}$$

Proposition 3.2 (Neumaier [160]) *If* $\boldsymbol{A}, \boldsymbol{B} \in \mathbb{IR}^{n\times n}$, *then*

$$(\boldsymbol{A} \pm \boldsymbol{B})^c = A^c + B^c, \tag{3.13}$$

$$(\boldsymbol{AB})^c = A^c B^c \text{ if } \boldsymbol{A} \text{ or } B^\Delta = 0, \tag{3.14}$$

$$A^\Delta = |\boldsymbol{A} - A^c| \leqslant |\boldsymbol{A}|, \tag{3.15}$$

$$(\boldsymbol{A} \pm \boldsymbol{B})^\Delta = A^\Delta + B^\Delta, \tag{3.16}$$

$$A^\Delta|\boldsymbol{B}| \leqslant (\boldsymbol{AB})^\Delta \leqslant A^\Delta|\boldsymbol{B}| + |A^c|B^\Delta, \tag{3.17}$$

$$|\boldsymbol{A}|B^\Delta \leqslant (\boldsymbol{AB})^\Delta \leqslant |\boldsymbol{A}|B^\Delta + A^\Delta|B^c|, \tag{3.18}$$

$$(\boldsymbol{AB})^\Delta = |\boldsymbol{A}|B^\Delta \text{ if } A^\Delta = 0 \text{ or } B^c = 0. \tag{3.19}$$

An $m \times n$ interval matrix can be interpreted as a set of real $m \times n$ matrices

$$\boldsymbol{A} = \left\{A \in \mathbb{R}^{m\times n} \mid A_{ij} \in \boldsymbol{A}_{ij}, i = 1, \ldots, m; j = 1, \ldots, n\right\}.$$

or it can be expressed equivalently as a matrix interval

$$\boldsymbol{A} = [\underline{A}, \overline{A}] = \{A \in \mathbb{R}^{m\times n} \mid \underline{A} \leqslant A \leqslant \overline{A}\}.$$

The inverse of an interval matrix is defined as

$$\boldsymbol{A}^{-1} \triangleq \text{hull}\left(\left\{A^{-1} \mid A \in \boldsymbol{A}\right\}\right). \tag{3.20}$$

Moreover, for a bounded set of real matrices $S = \{\tilde{A} \in \mathbb{R}^{m\times n}\}$, the $\inf(S)$ and $\sup(S)$ exist, and $\text{hull}(S) = [\inf(S), \sup(S)]$ is the tightest interval matrix that encloses set S.

For given positive vector $u \in \mathbb{R}^n$ (which is called a *scaling vector*), the *scaled maximum norm* is defined for an arbitrary interval vector $\boldsymbol{x} \in \mathbb{IR}^n$ and a square interval matrix $\boldsymbol{A} \in \mathbb{IR}^{n\times n}$ by

$$\|\boldsymbol{x}\|_u = \max\{|\boldsymbol{x}_i|/u_i \mid i = 1, 2, \ldots, n\}, \tag{3.21}$$

$$\|\boldsymbol{A}\|_u = \| |\boldsymbol{A}|u\| = \max\left\{\sum_{j=1}^{n} |\boldsymbol{A}_{ij}|u_j/u_i \mid i = 1, 2, \ldots, n\right\}. \tag{3.22}$$

It follows immediately from (3.21) and (3.22) that

$$\|\boldsymbol{x}\|_u \leqslant \alpha \Leftrightarrow |\boldsymbol{x}| \leqslant \alpha u,$$

$$\|\boldsymbol{x}\|_u < \alpha \Leftrightarrow |\boldsymbol{x}| < \alpha u,$$

$$\|\boldsymbol{A}\|_u \leqslant \alpha \Leftrightarrow |\boldsymbol{A}|u \leqslant \alpha u,$$

$$\|\boldsymbol{A}\|_u < \alpha \Leftrightarrow |\boldsymbol{A}|u < \alpha u.$$

Moreover, it is easy to verify that, for all $\boldsymbol{x} \in \mathbb{IR}^n$, $\boldsymbol{A}, \boldsymbol{B} \in \mathbb{IR}^{n\times n}$, the $\|\cdot\|_u$ norm satisfies the following rules:

$$\begin{aligned}
\|\boldsymbol{A}\|_u &= \||\boldsymbol{A}|\|_u,\\
\|\alpha\boldsymbol{A}\|_u &= |\alpha|\|\boldsymbol{A}\|_u \text{ for all } \alpha \in \mathbb{R},\\
\|\boldsymbol{A}\boldsymbol{x}\|_u &\leqslant \|\boldsymbol{A}\|_u\|\boldsymbol{x}\|_u,\\
\|\boldsymbol{A}\boldsymbol{B}\|_u &\leqslant \|\boldsymbol{A}\|_u\|\boldsymbol{B}\|_u,\\
|\boldsymbol{A}| \leqslant \boldsymbol{B} \Rightarrow \|\boldsymbol{A}\|_u &\leqslant \|\boldsymbol{B}\|_u,\\
\boldsymbol{B} \subseteq \boldsymbol{A} \Rightarrow \|\boldsymbol{B}\|_u &\leqslant \|\boldsymbol{A}\|_u.
\end{aligned}$$

Lemma 3.3 (Neumaier [160]) *Let $A \in \mathbb{R}^{n\times n}$ and $0 < u \in \mathbb{R}^n$. Then,*

$$|\lambda| \leqslant \|A\|_u \textit{ for every eigenvalue } \lambda \textit{ of } A. \tag{3.23}$$

Proof Let λ be an eigenvalue of A and x be an eigenvector associated with λ. Without loss of generality we can assume that $\|x\|_u = 1$. Then,

$$|\lambda| = |\lambda|\|x\|_u = \|\lambda x\|_u = \|Ax\|_u \leqslant \|A\|_u\|x\|_u = \|A\|_u.$$

□

The maximal absolute eigenvalue of a square matrix $A \in \mathbb{R}^{n\times n}$ is called the *spectral radius* and is denoted by $\rho(A)$. From Lemma 3.3, it follows straightforwardly that, for any $0 < u \in \mathbb{R}^n$,

$$\rho(A) \leqslant \|A\|_u.$$

Corollary 3.4 (Neumaier [160]) *If $A \in \mathbb{R}^{n\times n}$, $A \geqslant 0$ and $0 < \alpha \in \mathbb{R}$, then*

$$\rho(A) = \inf\{\|A\|_u \mid 0 < u \in \mathbb{R}^n\}, \tag{3.24}$$

$$\rho(A) < \alpha \Leftrightarrow \exists u > 0 : Au < \alpha u, \tag{3.25}$$

$$\rho(A) \geqslant \alpha \Leftrightarrow \exists u \geqslant 0 : Au \geqslant \alpha u \neq 0, \tag{3.26}$$

$$\rho(A) < 1 \Rightarrow \lim_{i\to+\infty} A^i = 0. \tag{3.27}$$

Proposition 3.5 (Neumaier [160]) *Let $A, B \in \mathbb{R}^{n\times n}$. Then,*

$$|A| \leqslant B \Rightarrow \rho(A) \leqslant \rho(|A|) \leqslant \rho(B), \tag{3.28}$$

$$\rho(AB) = \rho(BA). \tag{3.29}$$

Proposition 3.6 (Neumaier [160]) *If $A \in \mathbb{R}^{n\times n}$ has spectral radius $\rho(A) < 1$, then $I - A$ is non-singular. If additionally A is non-negative, then $(I - A)^{-1}$ is non-negative as well.*

Lemma 3.7 (Neumaier [160]) *If $P \in \mathbb{R}^{n\times n}$ is non-negative with $\rho(P) < \alpha$, then $\alpha I - P$ is non-singular, $(\alpha I - P)^{-1} \geqslant 0$, and $\rho((\alpha I - P)^{-1}) = 1/(\alpha - \rho(P))$.*

If P fulfills the conditions of Lemma 3.7, then by Corollary 3.4, there is a vector $u > 0$ such that $Pu \leqslant \alpha u$. Hence, matrix $A = (\alpha I - P)$ satisfies $Au > 0, A_{ij} \leqslant 0$ for $i \neq j$. This leads to the concept of an *M-matrix* [160].

Definition 3.8 An M-matrix is matrix $\mathbf{A} \in \mathbb{IR}^{n\times n}$ such that $\mathbf{A}_{ij} \leqslant 0$ for $i \neq j$ and $\mathbf{A}v > 0$ for some positive vector $v \in \mathbb{R}^n$.

Proposition 3.9 (Neumaier [160]) *Let $A, B \in \mathbb{R}^{n\times n}$, A be an M-matrix, and $B \geqslant 0$. Then,*

(i) *A is regular, $A_{ii} > 0$ for all i, and $A^{-1} \geqslant 0$,*
(ii) *$0 < x \in \mathbb{R}^n \Rightarrow A^{-1}x > 0$,*
(iii) *$A - B$ is an M-matrix iff $\rho(A^{-1}B) < 1$.*

Corollary 3.10 (Neumaier [160]) *Let $A \in \mathbb{R}^{n\times n}$ be such that $A_{ij} \leqslant 0$, $i \neq j$. Then, the following conditions are equivalent:*

(i) *A is an M-matrix,*
(ii) *$A^{-1} \leqslant 0$,*
(iii) *$u \geqslant 0, Au \leqslant 0 \Rightarrow u = 0$.*

Some of the properties presented in Proposition 3.9 and Corollary 3.10 extend to interval M-matrices. Before we present these properties, we must introduce the concepts of *regularity* and *strong regularity* of interval matrices.

Definition 3.11 A square interval matrix $\mathbf{A}$ is regular if every $A \in \mathbf{A}$ is non-singular.

Definition 3.12 Square interval matrix $\mathbf{A}$ is strongly regular if $(A^c)^{-1}\mathbf{A}$ is regular.

Remark The alternative definition of the strong regularity of interval matrices was given in [203] (cf. [17, 201, 218, 230]). It says that a square interval matrix $\mathbf{A}$ is strongly regular if A^c is non-singular and

$$\rho(|(A^c)^{-1}|A^{\Delta}) < 1. \tag{3.30}$$

However, as Theorem 3.18 shows, these two definitions of strong regularity are equivalent and, therefore, can be used interchangeably.

Theorem 3.13 (Neumaier [160])

(i) *If $\mathbf{A} \in \mathbb{IR}^{n\times n}$ is an M-matrix and $\mathbf{B} \subseteq \mathbf{A}$, then $\mathbf{B}$ is an M-matrix, in particular, all $A \in \mathbf{A}$ are M-matrices,*
(ii) *$\mathbf{A} \in \mathbb{IR}^{n\times n}$ is an M-matrix iff $\underline{A}$ and $\overline{A}$ are M-matrices,*
(iii) *Every M-matrix $\mathbf{A} \in \mathbb{IR}^{n\times n}$ is regular and*

$$\mathbf{A}^{-1} = [\overline{A}^{-1}, \underline{A}^{-1}] \geqslant 0. \tag{3.31}$$

M-matrices play an important role in the context of solving interval linear systems because of their unique properties, such as *positive inverse* property (3.31). Another important class of interval matrices are the so-called *H-matrices*, which are a generalization of M-matrices [160].

Definition 3.14 An H-matrix is matrix $\mathbf{A} \in \mathbb{IR}^{n\times n}$ such that its comparison matrix $\langle \mathbf{A} \rangle$ is an M-matrix.

Theorem 3.15 (Neumaier [160]) *Let $\mathbf{A} \in \mathbb{IR}^{n\times n}$ be regular.*

(i) If $(A^c)^{-1} \geqslant 0$, then $\mathbf{A}$ is strongly regular,
(ii) If A^c is an M-matrix, then $\mathbf{A}$ is an H-matrix.

Proposition 3.16 (Neumaier [160]) *For $\mathbf{A} \in \mathbb{IR}^{n\times n}$, the following conditions are equivalent:*

(i) $\mathbf{A}$ is an H-matrix,
(ii) $\langle \mathbf{A} \rangle$ is regular, and $\langle \mathbf{A} \rangle^{-1} u > 0$, where $u = (1, \dots, 1)^T$,
(iii) $0 \leqslant u \in \mathbb{R}^n$, $\langle \mathbf{A} \rangle u \leqslant 0 \Rightarrow u = 0$.

Theorem 3.13 extends for H-matrices in the following form.

Theorem 3.17 (Neumaier [160])

(i) If $\mathbf{A} \in \mathbb{IR}^{n\times n}$ is an H-matrix and $\mathbf{B} \subseteq \mathbf{A}$ then $\mathbf{B}$ is an H-matrix; in particular, all $A \in \mathbf{A}$ are H-matrices,
(ii) $\mathbf{A} \in \mathbb{IR}^{n\times n}$ is an H-matrix iff A^c is an H-matrix and

$$\rho(\langle A^c \rangle^{-1} A^{\Delta}) < 1; \tag{3.32}$$

in this case

$$\langle \mathbf{A} \rangle = \langle A^c \rangle - A^{\Delta}. \tag{3.33}$$

(iii) Every H-matrix $\mathbf{A} \in \mathbb{IR}^{n\times n}$ is regular and satisfies

$$|\mathbf{A}^{-1}| \leqslant \langle \mathbf{A} \rangle^{-1}, \tag{3.34}$$

with equality, e.g., if $\mathbf{A}$ is an M-matrix.

Theorem 3.18 (Neumaier [160]) *Let $\mathbf{A} \in \mathbb{IR}^{n\times n}$ and suppose that A^c is nonsingular. Then, the following conditions are equivalent:*

(i) $\mathbf{A}$ is strongly regular,
(ii) $\mathbf{A}^T$ is strongly regular,
(iii) $(A^c)^{-1}\mathbf{A}$ is regular,
(iv) $\rho(|(A^c)^{-1}|A^{\Delta}) < 1$,
(v) $\|I - (A^c)^{-1}\mathbf{A}\|_u < 1$ for some $u > 0$,
(vi) $(A^c)^{-1}\mathbf{A}$ is an H-matrix.

3.2 Parametric Interval Matrix

Real n-dimensional *parametric (parameter-dependent) matrix* $A(p)$ is a square array of elements that are real valued functions of K-dimensional vector of parameters $p = (p_1, \ldots, p_K) \in \mathbb{R}^K$, i.e., for each $i, j = 1, \ldots, n$,

$$A_{ij} : \mathbb{R}^K \ni (p_1, \ldots, p_K) \to A_{ij}(p_1, \ldots, p_K) \in \mathbb{R}. \tag{3.35}$$

If parameters $p_1, \ldots, p_K$ are uncertain and the only available information is that they vary within given intervals $\boldsymbol{p}_1, \ldots, \boldsymbol{p}_K$, a *parametric interval matrix* is obtained.

Definition 3.19 An $n \times n$ parametric interval matrix $A(\boldsymbol{p})$ with $\boldsymbol{p} \in \mathbb{R}^K$ is defined to be the following family of real parametric matrices:

$$A(\boldsymbol{p}) = \left\{A(p) \in \mathbb{R}^{n\times n} \mid p \in \boldsymbol{p}\right\}. \tag{3.36}$$

A parametric interval (column) vector can be thought of as a one-column parametric interval matrix.

Functions A_{ij} $(i, j = 1, \ldots, n)$ can be generally divided into affine-linear and nonlinear ones, therefore, we distinguish matrices with affine-linear dependencies and matrices with nonlinear dependencies. In the nonlinear case, we usually assume that functions A_{ij} are continuous and differentiable on $\boldsymbol{p}$. In practical computation, A_{ij} are often described by closed-form expressions.

If there are only affine-linear dependencies in $A(\boldsymbol{p})$, then $A(\boldsymbol{p})$ can be represented as

$$A(\boldsymbol{p}) = \left\{\left[a_{ij}^{(0)} + \sum_{k=1}^{K} a_{ij}^{(k)} p_k\right] \middle| p \in \boldsymbol{p}\right\}, \tag{3.37}$$

where $a_{ij}^{(k)} \in \mathbb{R}$ $(k = 0, \ldots, K,\ i, j = 1, \ldots, n)$, or equivalently (and more conveniently) as

$$A(\boldsymbol{p}) = \left\{A^{(0)} + \sum_{k=1}^{K} A^{(k)} p_k \,\middle|\, p \in \boldsymbol{p}\right\}, \tag{3.38}$$

where $A^{(k)} = \left[a_{ij}^{(k)}\right] \in \mathbb{R}^{n\times n}$ for $k = 0, \ldots, K$.

Affine-linear dependencies are not easy to handle, however, to deal with nonlinear dependencies, some sophisticated tools for bounding the range of multivariate functions on a box (see Sect. 1.3) are required in addition [259]. However, by using revised affine arithmetic, the nonlinear case can be reduced to an affine-linear one. Obviously, some loss of information is inevitable, but this is compensated for by the substantial simplification of a computation.

3.3 Affine Transformation

Let $A(\boldsymbol{p})$ be a parametric interval matrix with nonlinear dependencies and let $A(p)$ be an arbitrary matrix in $A(\boldsymbol{p})$. Each parameter $p_k \in \boldsymbol{p}_k$, $k = 1, \ldots, K$, can be represented by the revised affine from $\hat{p}_k = p_k^c + p_k^{\Delta}\varepsilon_k$, where ε_k is a noise symbol that varies independently from the other noise symbols within $\boldsymbol{U}$. So, we substitute $\hat{p}_k$ for p_k in $A(p)$ and perform the respective operations on the revised affine forms. As a result of the *affine transformation* of $A(\boldsymbol{p})$, we obtain the following matrix of revised affine forms (*revised affine matrix*):

$$\boldsymbol{C}(e) = C^{(0)} + \sum_{k=1}^{K} C^{(k)}\varepsilon_k + C^r[-1, 1] = \sum_{k=1}^{K} C^{(k)}\varepsilon_k + \boldsymbol{C}, \tag{3.39}$$

where C^r is a matrix of radii of the accumulative errors resulting from nonlinear operations (in rigorous computation also from rounding errors). For each $C(e) \in \boldsymbol{C}(e)$, there is a vector $e \in \boldsymbol{U}^K$ and $C \in \boldsymbol{C}$ such that $C(e) = C^{(0)} + \sum_{k=1}^{K} C^{(k)}\varepsilon_k$. Thus, revised affine matrix $\boldsymbol{C}(e)$ can be considered as an infinite family of real parametric matrices with affine-linear dependencies and $K + n^2$ normalized parameters, i.e., parameters varying within interval $[-1, 1]$.

Example 3.20 Consider the following parametric interval matrix with nonlinear dependencies:

$$A(\boldsymbol{p}) = \begin{pmatrix} p_1^2 & -2 & p_1p_2 \\ 1 & p_1 + 2p_2 & p_2 \\ p_2 & p_1p_2^2 & 0 \end{pmatrix},$$

where $p_1, p_2 \in [1, 2]$. The affine transformation of $A(\boldsymbol{p})$ yields matrix of revised affine forms $\boldsymbol{C}(e) = C(e) + \boldsymbol{C}$, where

$$C(e) = \begin{pmatrix} 1.5\varepsilon_1 & 0 & 0.75\varepsilon_1 + 0.75\varepsilon_2 \\ 0 & 0.5\varepsilon_1 + \varepsilon_2 & 0.5\varepsilon_2 \\ 0.5\varepsilon_2 & 1.125\varepsilon_1 + 2.25\varepsilon_2 & 0 \end{pmatrix},$$

$$\boldsymbol{C} = \begin{pmatrix} 2.375 & -2 & 2.25 \\ 1 & 4.5 & 1.5 \\ 1.5 & 3.729 & 0 \end{pmatrix} + \begin{pmatrix} 0.125 & 0 & 0.25 \\ 0 & 0 & 0 \\ 0 & 0.896 & 0 \end{pmatrix} \cdot [-1, 1].$$

Proposition 3.21 *Let $A(\boldsymbol{p})$ be a parametric interval matrix with nonlinear dependencies and let $\boldsymbol{C}(e)$ be its affine transformation. Then, $A(\boldsymbol{p}) \subseteq \boldsymbol{C}(e)$.*

Proof The inclusion follows directly from the fundamental invariant of affine arithmetic (FIAA) (see Sect. 2.1). □

If there are only affine-linear dependencies in $A(\boldsymbol{p})$, i.e., there is no interval part in revised affine matrix $\boldsymbol{C}(e)$, then we can write $\boldsymbol{C}(e)$ as ordinary parametric interval matrix $C(\boldsymbol{e})$.

Proposition 3.22 *Let $A(\boldsymbol{p})$ be a parametric interval matrix with affine-linear dependencies and let $C(\boldsymbol{e})$ be its affine transformation. Then $A(\boldsymbol{p}) = C(\boldsymbol{e})$.*

Proof If there are only linear dependencies in $A(\boldsymbol{p})$, then we have

$$\begin{aligned} A(\boldsymbol{p}) \ni A(p) &= A^{(0)} + \sum_{k=1}^{K} A^{(k)} p_k = A^{(0)} + \sum_{k=1}^{K} A^{(k)} (p_k^c + p_k^{\Delta} \varepsilon_k) = \\ &= A^{(0)} + \sum_{k=1}^{K} A^{(k)} p_k^c + \sum_{k=1}^{K} A^{(k)} p_k^{\Delta} \varepsilon_k = \\ &= C^{(0)} + \sum_{k=1}^{K} C^{(k)} \varepsilon_k = C(e) \in C(\boldsymbol{e}). \end{aligned}$$

□

The affine transformation of parametric interval matrix $A(\boldsymbol{p})$ with affine-linear dependencies into parametric interval matrix $C(\boldsymbol{e})$ will be referred to as *normalization*.

3.4 NP-Hard Problems Related to Parametric Interval Matrices

As it is well known, basic properties of real matrices such as non-singularity or positive definiteness can be verified in a polynomial time. This, however, does not apply to parametric interval matrices. Below, we summarize NP-hardness results concerning the selected properties of parametric interval matrices. The validity of these results is asserted by the respective results for interval matrices (see, e.g., [39, 164, 171, 211, 213]) and the fact that a parametric interval matrix can be considered as an interval matrix with n^2 parameters.

Theorem 3.23 *For square parametric interval matrix $A(\boldsymbol{p})$, each of the following problems is NP-hard:*

(i) check whether $A(\boldsymbol{p})$ is regular,
(ii) compute the radius of regularity

$$r(A(\boldsymbol{p})) = \min\{r \geqslant 0 \mid \exists p \in p^c + r[-p^{\Delta}, p^{\Delta}] \ : \ A(p) \text{ is singular}\}, \tag{3.40}$$

(iii) compute

$$\max\{\det(A(p)) \mid p \in \boldsymbol{p}\}, \tag{3.41}$$

(iv) check whether $\|A(p)\|_2 < 1$ for each $p \in \boldsymbol{p}$,
(v) check whether $A(\boldsymbol{p})$ is Hurwitz stable (even for symmetric parametric interval matrices),

(vi) compute the radius of stability

$$s(A(\boldsymbol{p})) = \min\{r \geqslant 0 \mid \exists p \in p^c + r[-p^\Delta, p^\Delta] \ : \ A(p) \text{ is unstable}\}, \qquad (3.42)$$

(vii) check whether $A(\boldsymbol{p})$ is Schur stable (even for symmetric parametric interval matrices),

(ix) given $\lambda \in \mathbb{R}$, decide whether $A(p)x = \lambda x$ for some $p \in \boldsymbol{p}$.[1]

In the next sections, we will discuss various approaches that enable us to circumvent the exponential complexity of some of the above-mentioned problems.

3.5 Regularity

In general, parametric interval matrix $A(\boldsymbol{p})$ has a certain property if all matrices $A(p) \in A(\boldsymbol{p})$ have that property. One of the most-important properties of a parametric interval matrix is its *regularity*. This property is especially important for the problem of solving parametric linear interval equations, but it is also important for the verification of several frequently-used properties of parametric interval matrices, such as positive definiteness, P-property, stability, and Schur stability (cf. [212, 221]).

Definition 3.24 Square parametric interval matrix $A(\boldsymbol{p})$ is regular if, for each $p \in \boldsymbol{p}$, matrix $A(p)$ is non-singular.

Definition 3.25 An $n \times n$ parametric interval matrix $A(\boldsymbol{p})$ is singular if there is $p \in \boldsymbol{p}$ such that matrix $A(p)$ is singular.

Regularity of a parametric interval matrix guarantees that the solution set of a parametric interval linear system is bounded. However, from Theorem 3.23, it follows that we cannot expect a polynomial-time algorithm to exist that checks the regularity of interval matrices (according to the belief that P $\neq$ NP, cf. [56]). Therefore, for practical computations, we need some verifiable sufficient conditions for both the regularity and singularity of parametric interval matrices. The more we have such conditions the better, since some of them may be better-suited for specific classes of parametric interval matrices than the others.

Below, we give some verifiable sufficient conditions for the regularity and singularity of parametric interval matrices (cf. [255]).

Proposition 3.26 (cf. [166]) *Let $A(\boldsymbol{p})$ be an $n \times n$ parametric interval matrix with affine-linear dependencies. If $A(\boldsymbol{p})$ is singular, then inequality*

$$\left|A\left(p^c\right)x\right| \leqslant B|x|, \qquad (3.43)$$

[1] It is worth mentioning that, for interval matrices, the analogue problem for eigenvectors can be solved in polynomial time ([17], Theorem 4.1).

where

$$B = \left(\sum_{k=1}^{K} |A^{(k)}| p_k^{\Delta} \right), \tag{3.44}$$

has a nontrivial solution.

Proof If $A(\boldsymbol{p})$ is singular, then there is a vector $x \neq 0$ such that $A(p)x = 0$ for some $p \in \boldsymbol{p}$. So,

$$\begin{aligned} |A(p^c)x| &= |(A(p) - A(p^c))x| \leqslant |(A(p) - A(p^c)||x| = \\ &= \left| \sum_{k=1}^{K} A^{(k)}(p_k - p_k^c) \right| |x| \leqslant \left(\sum_{k=1}^{K} |A^{(k)}||p_k - p_k^c| \right) |x| \leqslant \\ &\leqslant \left(\sum_{k=1}^{K} |A^{(k)}| p_k^{\Delta} \right) |x| = B|x|. \end{aligned}$$

□

As the following example shows, the opposite implication (which holds true for classical interval matrices) is not fulfilled by parametric interval matrices.

Example 3.27 Consider matrix $A(p) = A^{(1)} p_1$, where $p_1 \in [0.5, 1.5]$ and

$$A^{(1)} = \begin{pmatrix} 2 & 3 \\ 4 & 5 \end{pmatrix}.$$

It is not hard to see that inequality (3.43) is satisfied for $x = (-1, 1)^T$. But $\det(A(p)) = -2p_1 < 0$ for each $p_1 \in [0.5, 1.5]$, which means that $A(\boldsymbol{p})$ is regular.

Theorem 3.28 *Let $A(\boldsymbol{p})$ be a square parametric interval matrix with affine-linear dependencies such that $A(p^c)$ is non-singular, and let B be given by formula (3.44). If*

$$\rho \left(\left| A \left(p^c \right)^{-1} \right| B \right) < 1, \tag{3.45}$$

then $A(\boldsymbol{p})$ is regular.

Proof Suppose, to the contrary, that $A(\boldsymbol{p})$ is singular. Then, by Proposition 3.26, there is a nontrivial vector $x \in \mathbb{R}^n$ such that

$$|A(p^c)x| \leqslant B|x|.$$

Without loss of generality, we can assume that $x = A(p^c)^{-1}y$ for some $y \neq 0$. Hence,

$$|y| \leqslant B|A(p^c)^{-1}||y|,$$

and by Corollary 3.4 and Proposition 3.5

$$1 \leqslant \rho\left(B\left|A(p^c)^{-1}\right|\right) = \rho\left(\left|A(p^c)^{-1}\right|B\right),$$

which yields a contradiction. □

The opposite implication is not true in the general case. To show this, it is enough to consider the matrix from Example 3.27 (which is regular), whereas $\rho(|A(p^c)^{-1}|B) \approx 10.97$. In Sect. 3.6, we will discuss a class of matrices for which the condition from Theorem 3.28 is not only sufficient but also necessary.

The next two theorem (Theorems 3.29 and 3.30) generalize the results presented in [166, 202].

Theorem 3.29 *Let $A(\boldsymbol{p})$ be a square parametric interval matrix with affine-linear dependencies such that inequality*

$$\lambda_{\max}(B^T B) < \lambda_{\min}(A(p^c)^T A(p^c)) \tag{3.46}$$

holds true with B given by formula (3.44). Then, $A(\boldsymbol{p})$ is regular.

Proof Suppose to the contrary that $A(\boldsymbol{p})$ is singular. Then, by Proposition 3.26, there is a vector $x \neq 0$ such that

$$|A(p^c)x| \leqslant B|x|.$$

Without loss of generality, we can assume that $\|x\| = 1$ for some consistent norm $\|\cdot\|$. Since the smallest eigenvalue of a symmetric matrix $A(p^c)^T A(p^c)$ can be expressed as

$$\lambda_{\min}(A(p^c)^T A(p^c)) = \min_{\|x\|=1} \left\{x^T A(p^c)^T A(p^c)x\right\},$$

hence

$$\begin{aligned}\lambda_{\min}(A(p^c)^T A(p^c)) &\leqslant x^T A(p^c)^T A(p^c)x \leqslant |A(p^c)x|^T |A(p^c)x| \leqslant \\ &\leqslant (B|x|)^T B|x| \leqslant \lambda_{\max}(B^T B),\end{aligned}$$

which contradicts (3.46). □

Since $A(p^c)^T A(p^c)$ and $B^T B$ are both real symmetric matrices, we can find their eigenvalues using the iterative Jacobi eigenvalue algorithm [67, 89], which has $\mathcal{O}(n^3)$ asymptotic time complexity and quadratic convergence property. The computation of $A(p^c)^T A(p^c)$ and $B^T B$ can also be done in $\mathcal{O}(n^3)$ time, so the overall asymptotic time complexity of checking the regularity condition from Theorem 3.29 is $\mathcal{O}(n^3)$ (where n is the size of the matrix).

We can avoid the need to compute the eigenvalues in Theorem 3.29 if we instead use a positive definiteness check (cf. [202]).

Theorem 3.30 *Let $A(\boldsymbol{p})$ be a square parametric interval matrix with affine-linear dependencies. If matrix*

$$A(p^c)^T A(p^c) - \|B^T B\| I, \tag{3.47}$$

where B is given by formula (3.44) is positive definite for some consistent matrix norm $\|\cdot\|$, then $A(\boldsymbol{p})$ is regular.

Proof Suppose to the contrary that $A(\boldsymbol{p})$ is singular. Then, by Proposition 3.26, there is a nontrivial vector x satisfying (3.43). Without loss of generality, we can assume that $\|x\|_2 = 1$. Hence,

$$x^T A(p^c)^T A(p^c) x \leqslant \lambda_{\max}(B^T B) \leqslant \rho(B^T B) \leqslant \|B^T B\| = \|B^T B\|(x^T x).$$

Hence

$$x^T (A(p^c)^T A(p^c) - \|B^T B\| I) x \leqslant 0.$$

Thus, the matrix given by (3.47) is not positive definite, which yields a contradiction. □

Since $A(p^c)^T A(p^c) - \|B^T B\| I$ is a real symmetric matrix, we can use a modified Gaussian elimination [12] to check whether it is positive definite [202]. This follows from Sylvester's criterion, which states that a real symmetric matrix is positive definite if and only if all of its leading principal minors are positive. So, the asymptotic time computational of checking the regularity condition from Theorem 3.30 is $\mathcal{O}(n^3)$, where n is the size of the matrix. We can additionally decrease the number of operations by using the fact that Theorem 3.30 will remain valid if we replace $\|B^T B\|$ with $\|B^T\| \|B\|$ in (3.47). The regularity condition with $\|B^T\| \|B\|$ is weaker than (3.47), but we free ourselves from computing product $B^T B$ (cf. [202]).

Let $(\sigma_1, \sigma_2, \ldots)$ denote the singular values of a real matrix in descending order. In [230] Rump presented another sufficient condition for regularity of interval linear systems, which is used for large systems of linear and nonlinear equations. Similar condition can be formulated for parametric interval matrices.

Proposition 3.31 (cf. [230]) *Let $A(\boldsymbol{p})$ be a square parametric interval matrix with affine-linear dependencies such that*

$$\sigma_1(B) < \sigma_n(A(p^c)) \tag{3.48}$$

holds true, with B given by formula (3.44). Then, $A(\boldsymbol{p})$ is regular.

Proof The thesis of the theorem follows immediately from Theorem 3.29 and the fact that $\sigma_1^2(B) = \lambda_{\max}(B^T B)$ and $\sigma_n^2(A(p^c)) = \lambda_{\min}((A(p^c))^T A(p^c))$. □

Using the parametric interval matrix from Example 3.27 (which was proven to be regular), we can show that the opposite implication is not true since $3.67 \approx \sigma_1(B) > \sigma_n(A^c) \approx 0.27$.

The singular values of a real matrix can be obtained using Singular Value Decomposition (SVD). Given a real matrix A, its SVD decomposition is given by $U\Sigma V^T$, where Σ is a diagonal rectangular matrix with singular values of A on the diagonal. The overall time cost is $\mathcal{O}(mn^2)$ (cf. [66, 267]). Since $m = n$ in our case, the time cost of checking condition (3.48) is $\mathcal{O}(n^3)$.

3.6 Strong Regularity

As we have already mentioned, the regularity of parametric interval matrices is very important for the problem of solving parametric interval linear systems. In particular, the regularity of a parametric interval matrix is a sufficient condition for the solution set of a parametric interval linear system to be bounded. However, most of the methods for solving a parametric interval linear system (e.g., [80, 107, 113, 178, 234, 248]) require that the matrix of the system matrix is *strongly regular*. In this section, we will introduce the concept of the strong regularity of parametric interval matrices, and we will prove several sufficient and necessary strong regularity conditions (cf. [255]).

3.6.1 Pre-conditioning with Mid-Point Inverse

The pre-multiplication of a parametric interval matrix with a non-singular real matrix is referred to as *pre-conditioning*. Pre-conditioning with the inverse of the mid-point matrix is used most-often. This is motivated by the fact that, if p^{Δ} is small enough, then for each $p \in \boldsymbol{p}$, $A(p^c)^{-1}A(p) \approx I$, and the matrices close to the identity matrix are H-matrices and, hence, regular. Similar considerations for interval matrices [160] have led to the concept of *strongly regular* interval matrices. Obviously, strong regularity implies regularity (but not conversely).

In our recent paper [255], we introduced the definition of strong regularity, which was a straightforward generalization of Definition 3.12 (cf. [160]) to parametric interval matrices. This was motivated, among others, by the fact that most of the methods for solving parametric interval linear systems require that a pre-conditioned matrix is regular. Nevertheless, for the same reasons as the ones described above, we can also use *post-conditioning*, i.e., post-multiply $A(\boldsymbol{p})$ by the midpoint inverse. Obviously, each of these two transformations leads to a different class of parametric matrices that can be proven to be regular this way (cf. Example 3.38). Therefore, we use a definition in this book[2] that is consistent with the one given in [175].

---

[2]For the purposes of the methods described in this book, the definition from [255] is enough. However, in order to be up-to-date with the most-recent research on solving parametric interval linear systems, we adopt a slightly modified definition of strong regularity here. We would like, however, underline that post-conditioning so far has little practical use.

Definition 3.32 Square parametric interval matrix $A(\boldsymbol{p})$ is strongly regular if $A(p^c)$ is non-singular and at least one of the matrices

$$\boldsymbol{B} = \text{hull}\left(\{A(p^c)^{-1}A(p) \mid p \in \boldsymbol{p}\}\right), \tag{3.49a}$$

$$\boldsymbol{B}' = \text{hull}\left(\{A(p)A(p^c)^{-1} \mid p \in \boldsymbol{p}\}\right) \tag{3.49b}$$

is regular.

The next theorem can be used to compute $\boldsymbol{B}$ and $\boldsymbol{B}'$ if there are only affine-linear dependencies in $A(\boldsymbol{p})$.

Theorem 3.33 *Let $A(\boldsymbol{p})$ be a parametric interval matrix with affine-linear dependencies, let R, R' be some matrices, and let $\boldsymbol{C} = \text{hull}\left(RA(p)R' \mid p \in \boldsymbol{p}\right)$. Then,*

(i) $\boldsymbol{C} = RA^{(0)}R' + \sum_{k=1}^{K}\left(RA^{(k)}R'\right)\boldsymbol{p}_k$,
(ii) $C^c = RA(p^c)R'$, $C^{\Delta} = \sum_{k=1}^{K}\left|RA^{(k)}R'\right|p_k^{\Delta}$.

Proof

$$\begin{aligned}\boldsymbol{C} &= \text{hull}\left(\{RA(p)R' \mid p \in \boldsymbol{p}\}\right) = \\ &= \text{hull}\left(\left\{R\left(A^{(0)} + \sum_{k=1}^{K} A^{(k)}p_k\right)R' \;\middle|\; p \in \boldsymbol{p}\right\}\right) = \\ &= \text{hull}\left(\left\{RA^{(0)}R' + \sum_{k=1}^{K}\left(RA^{(k)}R'\right)p_k \;\middle|\; p \in \boldsymbol{p}\right\}\right) = \\ &= RA^{(0)}R' + \sum_{k=1}^{K}\left(RA^{(k)}R'\right)\boldsymbol{p}_k.\end{aligned}$$

The last equality holds, since each parameter p_k occurs only once in the expression.
Now, each parameter $\boldsymbol{p}_k$ can be written as $\boldsymbol{p}_k = p_k^c + [-p_k^{\Delta}, p_k^{\Delta}]$. Thus,

$$\begin{aligned}\boldsymbol{C} &= RA^{(0)}R' + \sum_{k=1}^{K}\left(RA^{(k)}R'\right)\left(p_k^c + [-p_k^{\Delta}, p_k^{\Delta}]\right) = \\ &= R\left(A^{(0)} + \sum_{k=1}^{K} A^{(k)}p_k^c\right)R' + \sum_{k=1}^{K}\left(RA^{(k)}R'\right)[-p_k^{\Delta}, p_k^{\Delta}] = \\ &= RA(p^c)R' + \left[-\sum_{k=1}^{K}\left|RA^{(k)}R'\right|p_k^{\Delta}, \sum_{k=1}^{K}\left|RA^{(k)}R'\right|p_k^{\Delta}\right],\end{aligned}$$

which completes the proof. □

Corollary 3.34 *Let $A(\boldsymbol{p})$ be a square parametric interval matrix with affine-linear dependencies such that $A(p^c)$ is non-singular, and let $\boldsymbol{B}$ be given by (3.49a). Then,*

(i) $\boldsymbol{B} = A(p^c)^{-1}A^{(0)} + \sum_{k=1}^{K}\left(A(p^c)^{-1}A^{(k)}\right)\boldsymbol{p}_k$,
(ii) $B^c = I$, $B^{\Delta} = \sum_{k=1}^{K}\left|A(p^c)^{-1}A^{(k)}\right|p_k^{\Delta}$.

Proof The proof of the theorem follows directly from Theorem 3.33 with $R = A(p^c)^{-1}$ and $R' = I$. □

Corollary 3.35 *Let $A(\boldsymbol{p})$ be a square parametric interval matrix with affine-linear dependencies such that $A(p^c)$ is non-singular, and let $\boldsymbol{B}'$ be given by (3.49b). Then,*

(i) $\boldsymbol{B}' = A^{(0)}A(p^c)^{-1} + \sum_{k=1}^{K} \left(A^{(k)}A(p^c)^{-1}\right)\boldsymbol{p}_k$,
(ii) $(B')^c = I$, $(B')^\Delta = \sum_{k=1}^{K} \left|A^{(k)}A(p^c)^{-1}\right| p_k^\Delta$.

Proof The proof of the theorem follows directly from Theorem 3.33 with $R = I$ and $R' = A(p^c)^{-1}$. □

The next theorem provides several necessary and sufficient conditions for the strong regularity of parametric interval matrices. This theorem extends Theorem 3.18 (cf. Theorem 1, [175]) and also our own result presented in [255].

Theorem 3.36 *Let $A(\boldsymbol{p})$ be a square parametric interval matrix with affine-linear dependencies, and suppose that $A(p^c)$ is non-singular. Let $\boldsymbol{B}$ be given by (3.49a) and $\boldsymbol{B}'$ be given by (3.49b). Then, the following conditions are equivalent:*

(i) $A(\boldsymbol{p})$ *is strongly regular,*
(ii) $\boldsymbol{B}$ *is regular or* $\boldsymbol{B}'$ *is regular,*
(iii) $|u| \leqslant B^\Delta|u|$ *has only trivial solution or* $|v| \leqslant (B')^\Delta|v|$ *has only one trivial solution,*
(iv) $\rho\left(B^\Delta\right) < 1$ *or* $\rho\left((B')^\Delta\right) < 1$,
(v) $B^\Delta u < u$, *for some* $u > 0$ *or* $(B')^\Delta v < v$, *for some* $v > 0$
(vi) $\left\|B^\Delta\right\|_u < 1$, *for some* $u > 0$ *or* $\left\|(B')^\Delta\right\|_v < 1$, *for some* $v > 0$,
(vii) $I - B^\Delta$ *is an* M*-matrix or* $I - (B')^\Delta$ *is an* M*-matrix,*
(viii) $\boldsymbol{B}$ *is an* H*-matrix or* $\boldsymbol{B}'$ *is an* H*-matrix.*

Theorem 3.36 presents several precise sufficient and necessary conditions under which the parametric interval matrix is strongly regular. These conditions are sufficient for the regularity of parametric interval matrices; however, as the following example shows, they are not necessary.

Example 3.37 Consider a parametric interval matrix defined by

$$A(p) = \begin{pmatrix} p_1 & -43 & 49 \\ -31 & p_1 & -35 \\ 25 & -35 & p_2 \end{pmatrix},$$

with $p_1 \in [31, 41]$, $p_2 \in [28, 38]$. Since $\det(A(p)) > 0$ (the minimum value of the determinant is 84; it is attained at the point (41, 28)), then $A(\boldsymbol{p})$ is regular, whereas $\rho(B^\Delta) = \rho((B')^\Delta) \approx 1.72$; thus, by Proposition 3.16, neither $\boldsymbol{B}$ nor $\boldsymbol{B}'$ is an H-matrix.

The next example (see [185]) shows that the class of parametric interval matrices that can be proven to be regular by using pre-conditioning differs from the class of parametric interval matrices that can be proven to be regular by using post-conditioning.

Example 3.38 ([185]) Consider the following three-dimensional parametric matrix:

$$A(p) = \begin{pmatrix} \frac{1}{2} - p_2 & p_1 & p_1 \\ p_2 & -p_2 & p_3 \\ p_1 & p_3 & 1 \end{pmatrix},$$

where $p_1 \in [3/4, 5/4]$, $p_2, p_3 \in [1/2, 3/2]$. Using a simple computation, we obtain that $\rho(B^{\Delta}) \approx 1.119$, whereas $\rho((B')^{\Delta}) \approx 0.969$ ($\|\boldsymbol{B}'\|_{(27,48,30)} \approx 0.98$), which means that $\boldsymbol{B}$ is not an H-matrix but $\boldsymbol{B}'$ is; thus, it is regular, which gives the strong regularity, and thus also, the regularity of $A(\boldsymbol{p})$.

Proposition 3.39 (cf. [202]) *Let $A(\boldsymbol{p})$ be a square parametric interval matrix with affine-linear dependencies such that $A(p^c)$ is non-singular, and let $\boldsymbol{B}$ and $\boldsymbol{B}'$ be given by (3.49a) and (3.49b), respectively. Then, the following two conditions are equivalent:*

(i) $(1 - \|(B^{\Delta})^T B^{\Delta}\|_u)I$ *is positive definite for some* $u > 0$ *or* $(1 - \|((B')^{\Delta})^T (B')^{\Delta}\|_v)I$ *is positive definite for some* $v > 0$,
(ii) $\lambda_{\max}((B^{\Delta})^T B^{\Delta}) < 1$ *or* $\lambda_{\max}(((B')^{\Delta})^T (B')^{\Delta}) < 1$.

If either of these conditions is satisfied, then $A(\boldsymbol{p})$ is strongly regular.

Proof The implication $(i) \Rightarrow (ii)$ is obvious.

$(ii) \Rightarrow (i)$ Assume that $\lambda_{\max}((B^{\Delta})^T B^{\Delta}) < 1$. Since $\rho((B^{\Delta})^T B^{\Delta}) = \lambda_{\max}((B^{\Delta})^T B^{\Delta}) < 1$, hence, by Corollary 3.4, there is a vector $u>0$ such that $\left((B^{\Delta})^T B^{\Delta}\right) u<u$. This means that $\|(B^{\Delta})^T B^{\Delta}\|_u < 1$ and, hence, matrix $(1 - \|(B^{\Delta})^T B^{\Delta}\|_u)I$ is positive definite. Similarly, we can show that, if $\lambda_{\max}(((B')^{\Delta})^T (B')^{\Delta}) < 1$, then matrix $(1 - \|((B')^{\Delta})^T (B')^{\Delta}\|_v)I$ is positive definite.

To prove the last assertion, suppose to the contrary that $A(\boldsymbol{p})$ is not strongly regular. Then, according to Theorem 3.36, there is a vector $u \neq 0$ such that $|u| \leqslant B^{\Delta}|u|$ and a vector $v \neq 0$ such that $|v| \leqslant (B')^{\Delta}|v|$. Without loss of generality, we can assume that $\|u\|_2 = 1$ and $\|v\|_2 = 1$. So,

$$1 = u^T u \leqslant |u|^T |u| \leqslant |u|^T (B^{\Delta})^T B^{\Delta} |u| \leqslant \lambda_{\max}((B^{\Delta})^T B^{\Delta})$$

and

$$1 = v^T v \leqslant |v|^T |v| \leqslant |v|^T ((B')^{\Delta})^T (B')^{\Delta} |v| \leqslant \lambda_{\max}(((B')^{\Delta})^T (B')^{\Delta}),$$

which contradicts (ii); hence, (i). □

The conditions from Proposition 3.39 are only sufficient conditions for the strong regularity of parametric interval matrices. This is illustrated by the following example.

Example 3.40 Consider parametric interval matrix

$$A(p) = \begin{pmatrix} 1 & 0 \\ 0 & 1 \end{pmatrix} + \begin{pmatrix} 0.5 & 1 \\ 0 & 0.5 \end{pmatrix} p_1, \ \boldsymbol{p}_1 \in [-1, 1].$$

In this case, $B^{\Delta} = (B')^{\Delta}$, so it is enough to consider only one of these two matrices. Since $\rho\left(B^{\Delta}\right) = 0.5 < 1$, $\boldsymbol{B}$ is regular; hence, $A(\boldsymbol{p})$ is strongly regular. However, $\|(B^{\Delta})^T B^{\Delta}\|_u \geqslant \lambda_{\max}((B^{\Delta})^T B^{\Delta}) \approx 1.46 > 1$.

3.6.2 Pre-conditioning with Arbitrary Non-singular Matrix

Due to the presence of rounding errors (which are inherent to floating-point arithmetic), it is difficult to obtain the exact inverse of mid-point matrix $A(p^c)$ in numerical computations. However, as we will show below, using the mid-point inverse is not mandatory. Instead, we can use any other real matrix R as a conditioning matrix. We can even combine both pre- and post-conditioning [160], providing that the resulting matrix fulfills some regularity conditions.

Theorem 3.41 *Let $A(\boldsymbol{p})$ be a parametric interval matrix with affine-linear dependencies, and let $\boldsymbol{C} = \text{hull}\left(\{RA(p) \mid p \in \boldsymbol{p}\}\right)$, $\boldsymbol{C}' = \text{hull}\left(\{A(p)R \mid p \in \boldsymbol{p}\}\right)$. Then, the following conditions are equivalent:*

(i) $A(\boldsymbol{p})$ is strongly regular,
(ii) $\boldsymbol{C}$ is an H-matrix or $\boldsymbol{C}'$ is an H-matrix for some $R \in \mathbb{R}^{n\times n}$,
(iii) C^c is an H-matrix and $\rho\left(\left|(C^c)^{-1}\right| C^{\Delta}\right) < 1$ or $(C')^c$ is an H-matrix and $\rho\left((C')^{\Delta}\left|((C')^c)^{-1}\right|\right) < 1$.

Proof (i)⇒(ii) If $A(\boldsymbol{p})$ is strongly regular, then $A(p^c)$ is regular and it is enough to put $R = A(p^c)^{-1}$.

(ii)⇒(iii) If $\boldsymbol{C}$ is an H-matrix, then C^c is an H-matrix as well and $\rho\left(\langle C^c\rangle^{-1} C^{\Delta}\right) < 1$ by Theorem 3.17. Now, since $\left|(C^c)^{-1}\right| \leqslant \langle C^c\rangle^{-1}$ by Theorem 3.17, $\left|(C^c)^{-1}\right| C^{\Delta} \leqslant \langle C^c\rangle^{-1} C^{\Delta}$; hence, $\rho\left(\left|(C^c)^{-1}\right| C^{\Delta}\right) \leqslant \rho\left(\langle C^c\rangle^{-1} C^{\Delta}\right) < 1$. Similarly, we can show that, if $\boldsymbol{C}'$ is an H-matrix, then $(C')^c$ is an H-matrix and $\rho\left((C')^{\Delta}\left|((C')^c)^{-1}\right|\right) < 1$.

(iii)⇒(i) If $C^c = RA(p^c)$ is an H-matrix, then it is regular; therefore, R and $A(p^c)$ are regular as well. Now, with $\boldsymbol{B}$ given by (3.49a), it holds that

$$\begin{aligned} B^{\Delta} &= \textstyle\sum_{k=1}^{K} \left|A(p^c)^{-1}A^{(k)}\right| p_k^{\Delta} = \sum_{k=1}^{K} \left|(RA(p^c))^{-1}RA^{(k)}\right| p_k^{\Delta} \leqslant \\ &\leqslant \left|(RA(p^c))^{-1}\right| \textstyle\sum_{k=1}^{K} \left|RA^{(k)}\right| p_k^{\Delta}. \end{aligned} \tag{3.50}$$

So, $\rho\left(B^{\Delta}\right) \leqslant \rho\left(\left|(C^c)^{-1}\right| C^{\Delta}\right) < 1$, and by Theorem 3.36, $A(\boldsymbol{p})$ is strongly regular. Similarly, we can show that, if $(C')^c$ is an H-matrix and $\rho\left((C')^{\Delta}\left|((C')^c)^{-1}\right|\right) < 1$, then $\rho\left((B')^{\Delta}\right) \leqslant \rho\left((C')^{\Delta}\left|((C')^c)^{-1}\right|\right) < 1$; thus, $A(\boldsymbol{p})$ is strongly regular. □

Inequalities

$$\rho\left(B^{\Delta}\right) \leqslant \rho\left(\left|(C^{c})^{-1}\right| C^{\Delta}\right),$$
$$\rho\left((B')^{\Delta}\right) \leqslant \rho\left((C')^{\Delta}\left|((C')^{c})^{-1}\right|\right)$$

indicate that, from a certain point of view (cf. [83]), the best choice for R is the inverse of the mid-point matrix. Therefore, in numerical computations, it is recommended to use the numerically computed mid-point inverse.

Another sufficient condition for the strong regularity of parametric interval matrices is strictly connected with Rump's iterative method for solving systems of equations [229]. Let us note that the condition presented in the next theorem can be applied to parametric matrices with arbitrary dependencies.

Theorem 3.42 *Let $A(\boldsymbol{p})$ be a parametric interval matrix, let R some matrix, and let $\boldsymbol{C} = \text{hull}\,(\{RA(p) \mid p \in \boldsymbol{p}\})$, $\boldsymbol{C}' = \text{hull}\,(\{A(p)R \mid p \in \boldsymbol{p}\})$. If*

$$\rho(|I - \boldsymbol{C}|) < 1 \quad \text{or} \quad \rho\left(\left|I - \boldsymbol{C}'\right|\right) < 1, \tag{3.51}$$

then $A(\boldsymbol{p})$ is strongly regular.

Proof If $\rho(|I - \boldsymbol{C}|) < 1$, then $|I - \boldsymbol{C}|u < u$ for some positive vector $u \in \mathbb{R}^n$. Since, by Proposition 3.1, $I - \langle \boldsymbol{C} \rangle \leqslant |I - \boldsymbol{C}|$, then $u - \langle \boldsymbol{C} \rangle u < u$, which gives us $\langle \boldsymbol{C} \rangle u > 0$. So, $\boldsymbol{C}$ is an H-matrix, and thus, by Theorem 3.41, $A(\boldsymbol{p})$ is strongly regular. Similarly, we can show that, if $\rho(|I - \boldsymbol{C}'|) < 1$, then $A(\boldsymbol{p})$ is strongly regular. □

In the affine-linear case, the interval computations (which are more time-consuming than ordinary floating-point computations) can be avoided if the following matrix is used for strong regularity verification purposes instead of matrix $|I - \boldsymbol{C}|$:

$$C = \left|I - RA(p^c)\right| + \textstyle\sum_{k=1}^{K} \left|RA^{(k)}\right| p_k^{\Delta} \tag{3.52}$$

Analogously, instead of matrix $|I - \boldsymbol{C}'|$, we can use matrix

$$C' = \left|I - A(p^c)R\right| + \textstyle\sum_{k=1}^{K} \left|A^{(k)}R\right| p_k^{\Delta}. \tag{3.53}$$

Corollary 3.43 *Let $A(\boldsymbol{p})$ be a parametric interval matrix with affine-linear dependencies and C, C' be given by (3.52) and (3.53), respectively, where R is some matrix. If*

$$\rho\left(C\right) < 1 \tag{3.54a}$$

or

$$\rho\left(C'\right) < 1, \tag{3.54b}$$

then $A(\boldsymbol{p})$ is strongly regular.

Proof Put $\boldsymbol{C} = \text{hull}\left(\{RA(p) \mid p \in \boldsymbol{p}\}\right)$. Then,

$$|I - \boldsymbol{C}| = \left|I - RA(p^c) + \left[-\sum_{k=1}^{K} \left|RA^{(k)}\right| p_k^{\Delta}, \sum_{k=1}^{K} |RA^{(k)}| p_k^{\Delta}\right]\right| =$$
$$= \left|I - RA\left(p^c\right)\right| + \sum_{k=1}^{K} \left|RA^{(k)}\right| p_k^{\Delta} = C.$$

Hence, $\rho(|I - \boldsymbol{C}|) = \rho(C) < 1$, and, by Theorem 3.42, $A(\boldsymbol{p})$ is strongly regular. Similarly, we can show that, if $\rho\left(C'\right) < 1$, then $A(\boldsymbol{p})$ is strongly regular. □

The next theorem shows that the best choice for R in (3.54a) is the mid-point inverse. This theorem is a generalization of the result from [201].

Theorem 3.44 *Let $A(\boldsymbol{p})$ be a parametric interval matrix with affine-linear dependencies. Let $\rho(C) < 1$, where C is given by (3.52), hold for some R. Then, $A(p^c)$ is non-singular and it holds that*

$$\rho(B^{\Delta}) \leqslant \rho(C), \tag{3.55}$$

where $\boldsymbol{B}$ is given by (3.49a).

Proof Since $I - RA(p^c) \leqslant |I - RA(p^c)| \leqslant C$, $\rho(I - RA(p^c)) \leqslant \rho(C) < 1$. Then, matrix $RA(p^c) = I - (I - RA(p^c))$ is non-singular, and so is $A(p^c)$.

Now, suppose that $\rho\left(B^{\Delta}\right) > \rho(C)$. Then, $\rho(C) < \beta = \min\left\{1, \rho\left(B^{\Delta}\right)\right\}$; and, for $\alpha = \left(\rho(C) + \beta\right)/2$, it holds that $Cx < \alpha x$ for some $x > 0$. Since $\alpha < 1$, then

$$\alpha\left|I - RA(p^c)\right| x + \left(\sum_{k=1}^{K} \left|RA^{(k)}\right| p_k^{\Delta}\right) x < \alpha x, \tag{3.56}$$

is also true, which implies

$$\left(\sum_{k=1}^{K} \left|RA^{(k)}\right| p_k^{\Delta}\right) x < \alpha(I - |I - RA(p^c)|)x. \tag{3.57}$$

Because $\rho\left(|I - RA(p^c)|\right) < 1$, matrix $I - |I - RA(p^c)|$ is nonnegative invertible. Pre-multiplying (3.57) by its inverse gives

$$\left(I - \left|I - RA(p^c)\right|\right)^{-1} \left(\sum_{k=1}^{K} \left|RA^{(k)}\right| p_k^{\Delta}\right) x < \alpha x. \tag{3.58}$$

Now,

$$B^{\Delta} = \sum_{k=1}^{K} \left|A(p^c)^{-1} A^{(k)}\right| p_k^{\Delta} = \tag{3.59}$$
$$= \sum_{k=1}^{K} \left|(RA(p^c))^{-1} RA^{(k)}\right| p_k^{\Delta} \leqslant \left|(RA(p^c))^{-1}\right| \sum_{k=1}^{K} \left|RA^{(k)}\right| p_k^{\Delta}.$$

Since $\rho(I - RA(p^c)) < 1$, the inverse of $RA(p^c)$ can be expressed by a Neumann series, i.e., $(RA(p^c))^{-1} = \sum_{j=0}^{\infty} (I - RA(p^c))^j$. So,

$$\left|(RA(p^c))^{-1}\right| \leqslant \sum_{j=0}^{\infty} |I - RA(p^c)|^j = (I - |I - RA(p^c)|)^{-1}. \tag{3.60}$$

From (3.59) and (3.60), it follows that

$$B^{\Delta} \leqslant (I - |I - RA(p^c)|)^{-1} \sum_{k=1}^{K} |RA^{(k)}| p_k^{\Delta}.$$

Since $x > 0$,

$$B^{\Delta} x \leqslant (I - |I - RA(p^c)|)^{-1} \left(\sum_{k=1}^{K} \left|RA^{(k)}\right| p_k^{\Delta} \right) x,$$

and, in view of (3.58), $B^{\Delta} x \leqslant \alpha x$. This implies contradiction $\rho\left(B^{\Delta}\right) \leqslant \alpha < \rho\left(B^{\Delta}\right)$. Therefore, (3.55) holds, which completes the proof. □

Corollary 3.45 *Let $A(\boldsymbol{p})$ be a parametric interval matrix with affine-linear dependencies. Let $\rho(C') < 1$, where C' is given by (3.53), hold for some matrix R. Then, $A(p^c)$ is non-singular, and*

$$\rho((B')^{\Delta}) \leqslant \rho(C'), \tag{3.61}$$

where $\boldsymbol{B}'$ is given by (3.49b).

Proof We have

$$C'^{T} = \left|I - R^{T} \left(A(p^c)\right)^{T}\right| + \sum_{k=1}^{K} \left|R^{T} \left(A^{(k)}\right)^{T}\right| p_k^{\Delta}.$$

Since $\rho\left(C'^{T}\right) = \rho(C') < 1$, hence, by Theorem 3.44, $A(p^c)^T$ is non-singular, so $A(p^c)$ is non-singular as well. Moreover, it holds that

$$\rho\left((B')^{\Delta}\right) = \rho\left(\left((B')^{\Delta}\right)^{T}\right) \leqslant \rho\left(\left(C'\right)^{T}\right) = \rho(C'). \qquad \square$$

Conditions (3.51) and (3.54) are only sufficient conditions for the strong regularity of parametric interval matrices, as illustrated by the following example.

Example 3.46 Consider a parametric interval matrix defined by $\boldsymbol{p} = [1, 3]$,

$$A(p) = \begin{pmatrix} 1 & 0 \\ 0 & 1 \end{pmatrix} p, \quad R = \begin{pmatrix} 2 & 0 \\ 0 & 2 \end{pmatrix}.$$

Since $\boldsymbol{C} = \boldsymbol{C}'$ and $C = C'$, we can restrict ourselves to $\boldsymbol{C}$ and C. Using a simple computation, we obtain that, for $u = (1, 1)^T > 0$,

$$\langle \boldsymbol{C} \rangle u = \begin{pmatrix} 2 \\ 2 \end{pmatrix} > 0,$$

which means that $\boldsymbol{C}$ is an H-matrix; thus, $A(\boldsymbol{p})$ is strongly regular by Theorem 3.41, whereas $\rho(C) = (|I - \boldsymbol{C}|) = 5 > 1$.

3.7 Radius of Regularity

Parametric interval matrix $A(\boldsymbol{p})$ with affine-linear dependencies can be normalized by using the affine transformation from Sect. 3.3. So, without loss of generality, we can assume that $\boldsymbol{p} = [-1, 1]^K$. Then, formula (3.40) takes the following form (cf. [114]):

$$r(A(\boldsymbol{p})) = \min\{r \geqslant 0 \mid \exists p \in \boldsymbol{p} \ : \ A(rp) \text{ is singular}\}. \tag{3.62}$$

The value of $r(A(\boldsymbol{p}))$ provides us a quantitative measure of a distance from singularity [111]. Obviously, if $r(A(\boldsymbol{p})) > 1$, then $A(\boldsymbol{p})$ is regular. Conversely, if $r(A(\boldsymbol{p})) \leqslant 1$, then $A(\boldsymbol{p})$ is singular. But, as is already known, the problem of computing the radius of the regularity is, in general, NP-hard. An interesting approach to circumventing the exponential complexity was proposed by Kolev [114] (cf. [111]). Below, we present the general idea of this approach, and we provide a new proof of the main result.

It is not hard to see that $r(A(\boldsymbol{p})) = 0$ if and only if $A(p^c) = A^{(0)}$ is singular. Since the case of $A^{(0)}$ is trivial, we can assume that $A^{(0)}$ is non-singular in what follows, i.e., $r(A(\boldsymbol{p})) > 0$. Moreover, similarly as in [114], we assume that $r(A(\boldsymbol{p})) < +\infty$. Notice that verifying the first assumption is a trivial task, whereas the time complexity of verifying the second assumption is an open question.

Given $r > 0$, the singularity of matrix $A(rp)$ is equivalent to the existence of the solution to the following *generalized eigenvalue problem*

$$B(p)x = \lambda A_0 x, \tag{3.63}$$

where $B(p) = \sum_{k=1}^{K} A^{(k)} p_k$, $A_0 = -A^{(0)}$, $\lambda = 1/r$.

For arbitrary parametric matrix $A(p) \in A(\boldsymbol{p})$, we denote (cf. [171])

$$\rho_0(A(p)) = \max\left\{|\lambda| \mid B(p)x = \lambda A_0 x, x \neq 0, \lambda \in \mathbb{R}\right\}. \tag{3.64}$$

This is an analogue of the spectral radius, except that the maximum is taken over real eigenvalues. If no real eigenvalue exists, then $\rho_0(A(p))$ is set to zero. Similarly, for arbitrary parametric interval matrix $A(\boldsymbol{p})$, we denote

$$\rho_0(A(\boldsymbol{p})) = \max\{\rho_0(A(p)) \mid p \in \boldsymbol{p}\}, \tag{3.65}$$

i.e., the *real maximum magnitude* (RMM) eigenvalue of the following class of the generalized eigenvalue problems (cf. [114]):

$$B(p)x = \lambda A_0 x, \ \ p \in \boldsymbol{p}. \tag{3.66}$$

Theorem 3.47 (Kolev [114]) *Let $A(\boldsymbol{p})$ be a parametric interval matrix such that $A^{(0)}$ is non-singular and $r(A(\boldsymbol{p})) < +\infty$. Then,*

$$r(A(\boldsymbol{p})) = \frac{1}{\rho_0(A(\boldsymbol{p}))}. \tag{3.67}$$

Proof Since $A^{(0)}$ is non-singular, then $r(A(\boldsymbol{p})) > 0$. So, take arbitrary $r > 0$ and assume that $A(rp)$ is singular for some $p \in \boldsymbol{p}$. This is equivalent to the existence of the solution to the generalized eigenvalue problem (3.63) with $\lambda = 1/r$. Since λ is a positive real number, $\lambda \leqslant \max\{\rho_0(A(p)), p \in \boldsymbol{p}\}$, which is true if and only if $r \geqslant 1/\max\{\rho_0(A(p)), p \in \boldsymbol{p}\}$. So, the minimum value of r is given by (3.67). □

Hence, the problem of computing the radius of the regularity has been reduced to the problem of computing the RMM eigenvalue of (3.66). The analogous result presented in [111] indicates that this approach can be useful for elaborating a method for determining $r(A(\boldsymbol{p}))$ whose time complexity is not *a priori* exponential [114].

3.8 Radius of Strong Regularity

Formula (3.40) is a generalization of the following formula for the radius of the regularity of an interval matrix:

$$r_I(\boldsymbol{A}) = \min\{r \geqslant 0 \mid \exists A \in A^c + r[-A^\Delta, A^\Delta] \; : \; A \text{ is singular}\}. \tag{3.68}$$

Since $A(\boldsymbol{p})$ is strongly regular if at least one of the matrices $\boldsymbol{B}, \boldsymbol{B}'$ given by (3.49) is regular, we can define the *radius of strong regularity* of a parametric interval matrix $A(\boldsymbol{p})$ as

$$r^*(A(\boldsymbol{p})) = \max\{r_I(\boldsymbol{B}), r_I\left(\boldsymbol{B}'\right)\}. \tag{3.69}$$

Obviously, $r^*(A(\boldsymbol{p})) \leqslant r(A(\boldsymbol{p}))$.

Let $Q = \{-1, +1\}^n$ and $T_y = \operatorname{diag}(y)$. Based on the result presented in [171], we can formulate the following theorem:

Proposition 3.48 *Let $A(\boldsymbol{p})$ be a square parametric interval matrix with affine-linear dependencies such that $A(p^c)$ is non-singular. Then,*

$$r^*(A(\boldsymbol{p}) = \max\left\{\frac{1}{\rho(B^\Delta)}, \frac{1}{\rho((B')^\Delta)}\right\}. \tag{3.70}$$

Proof Since $B^c = I$, $(B^c)' = I$, and B^Δ, $\left(B^\Delta\right)' \geqslant 0$, hence (Theorem 2.1, [171])

$$\boldsymbol{r}_I(\boldsymbol{B}) = \frac{1}{\max\limits_{y,z\in Q} \rho_0(T_y B^\Delta T_z)} = \frac{1}{\rho\left(B^\Delta\right)},$$

$$\boldsymbol{r}_I((\boldsymbol{B})') = \frac{1}{\max\limits_{y,z\in Q} \rho_0(T_y B^\Delta T_z)} = \frac{1}{\rho\left((B^\Delta)\right)},$$

which, in view of (3.69), gives (3.70). □

3.9 Positive (Semi-)Definiteness

Positive (semi-)definiteness of parametric interval matrices is important, e.g., for B&B-based interval global optimization, which in turn can be employed to develop methods (see Sect. 5.3.1) for computing the hull of the parametric solution set (defined in Chap. 4). As is well known, the convexity of function f on box $\boldsymbol{x}$ can be verified by checking the positive definiteness of all Hessian matrices $\nabla^2 f(x)$ where $x \in \boldsymbol{x}$. Hence, if $\nabla^2 \boldsymbol{f}(\boldsymbol{x})$ is positive definite, then f is strictly convex on $\boldsymbol{x}$, and the search within the box becomes much easier. On the other hand, if the Hessian is neither positive nor negative definite on $\boldsymbol{x}$, then the box can be reduced or removed, as it contains no local (and hence, no global) minimum in $\boldsymbol{x}$. Positive definiteness can be useful as well when checking whether $f(x) \geqslant a$ for each $x \in \boldsymbol{x}$, where a is a prescribed value [38]. All consideration presented in this section will generally concern parametric interval matrices with affine-linear dependencies unless stated otherwise.

Definition 3.49 Parametric interval matrix $A(\boldsymbol{p})$ is (strongly) positive definite (semi-definite) if $A(p)$ is positive definite (semi-definite) for each $p \in \boldsymbol{p}$.

Definition 3.50 Parametric interval matrix $A(\boldsymbol{p})$ is weakly positive definite (semi-definite) if $A(p)$ is positive definite (semi-definite) for at least one $p \in \boldsymbol{p}$.

The problem of checking whether a parametric interval matrix is positive (semi-) definite is co-NP hard (cf. [127]). However, based on the following theorems, the positive definiteness of large parametric interval matrices can be verified in a reasonable amount of time, assuming the number of parameters is small enough [84].

Theorem 3.51 (Hladík [84]) *For square parametric interval matrix $A(\boldsymbol{p})$, the following assertions are equivalent:*

(i) *$A(\boldsymbol{p})$ is positive semi-definite,*
(ii) *$A(p)$ is positive semi-definite for each p such that $p_k \in \{\underline{p}_k, \overline{p}_k\}$, $k = 1, \ldots, K$ (vertex property),*
(iii) $x^T A(p^c)x - \sum_{k=1}^K \left|xA^{(k)}x\right| p_k^\Delta \geqslant 0$ *for each* $0 \neq x \in \mathbb{R}^n$.

Proof Implication (i)⇒(ii) is obvious. To prove (ii)⇒(iii), take arbitrary $x \neq 0$ and put $s_k = \operatorname{sgn}\left(x^T A^{(k)} x\right)$. Then

$$x^T A^c x - \sum_{k=1}^{K} \left|x^T A^{(k)} x\right| p_k^\Delta = x^T A^c x - \sum_{k=1}^{K} s_k x^T A^{(k)} x p_k^\Delta =$$
$$x^T \left(A^{(0)} + \sum_{k=1}^{K} A^{(k)} \left(p^c - s_k p_k^\Delta\right) \right) x \geqslant 0.$$

The last inequality follows from the fact that $p_k^c - s_k p_k^\Delta \in \{\underline{p}_k, \overline{p}_k\}$, $k = 1, \ldots, K$.

To prove (iii)⇒(i), take arbitrary $x \neq 0$ and $p \in \boldsymbol{p}$. Then,

$$x^T A(p)x = x^T A^c x + \sum_{k=1}^{K} x^T A^{(k)} x\,(p_k - p_k^c) \geqslant$$
$$\geqslant x^T A^c x - \sum_{k=1}^{K} \left|x^T A^{(k)} x\right| |p_k - p_k^c| \geqslant x^T A^c x - \sum_{k=1}^{K} \left|x^T A^{(k)} x\right| p_k^{\Delta} \geqslant 0.$$

□

According to the vertex property, in order to check the strong positive semi-definiteness of a parametric interval matrix, it is enough to consider 2^K real matrices. This number can be further decreased in some cases based on the following result (cf. [84]):

Theorem 3.52 (Hladík [84]) *For square parametric interval matrix $A(\boldsymbol{p})$, the following assertions hold true:*

(i) If, for some $k \in \{1, \ldots, K\}$, $A^{(k)}$ is positive semi-definite, then we can set $p_k = \underline{p}_k$ for checking the positive semi-definiteness,
(ii) If, for some $k \in \{1, \ldots, K\}$, $A^{(k)}$ is positive semi-negative, then we can set $p_k = \overline{p}_k$ for checking positive semi-definiteness.

Proof Let $p \in \boldsymbol{p}$ and let $A^{(k)}$ be positive semi-definite. To prove the first assertion, it is enough to show that if matrix

$$A^{(0)} + \sum_{i=1, i\neq k}^{K} A^{(i)} p_i + A^{(k)} \underline{p}_k$$

is positive semi-definite, then $A(p)$ is positive semi-definite as well. But, if $A^{(k)}$ is positive semi-definite, then $A^{(k)}(p_k - \underline{p}_k)$ is also positive semi-definite as well and since

$$A(p) = \left(A^{(0)} + \sum_{i=1, i\neq k}^{K} A^{(i)} p_i + A^{(k)} \underline{p}_k \right) + A^{(k)}(p_k - \underline{p}_k),$$

hence $A(p)$ is positive semi-definite as a sum of positive semi-definite matrices.

The second assertion can be proved in a similar manner. It is enough to observe that if $A^{(k)}$ is negative semi-definite, then $A^{(k)}(p_k - \overline{p}_k)$ is positive semi-definite. □

If the number of parameters is still too large, then the following sufficient condition can be used [84] (however, it only applies to symmetric parametric interval matrices). Before we proceed further, the definition of a symmetric parametric interval matrix must be introduced.

Definition 3.53 Square parametric interval matrix $A(\boldsymbol{p})$ is symmetric if, for each $k = 0, \ldots, K$, $A^{(k)}$ is symmetric.

Theorem 3.54 (Hladík [84]) *Let $A(\boldsymbol{p})$ be a symmetric parametric interval matrix. Let $A^{(k)} = A_1^{(k)} + A_2^{(k)}$ for each $k = 1, \ldots, K$, where $A_1^{(k)}$ is positive semi-definite and $A_2^{(k)}$ is negative semi-definite. If matrix*

$$A_{12} = A^{(0)} + \sum_{k=1}^{K} \left(A_1^{(k)} \underline{p}_k + A_2^{(k)} \overline{p}_k \right) \tag{3.71}$$

is positive semi-definite, then $A(\boldsymbol{p})$ is positive semi-definite.

Proof Let $p \in \boldsymbol{p}$. Since, for each $k = 1, \ldots, K$, $A^{(k)} = A_1^{(k)} + A_2^{(k)}$, hence

$$A(p) = A^{(0)} + \sum_{k=1}^{K} \left(A_1^{(k)} + A_2^{(k)} \right) p_k =$$

$$= \left(\sum_{k=1}^{K} A_1^{(k)} (p_k - \underline{p}_k) + \sum_{k=1}^{K} A_2^{(k)} (p_k - \overline{p}_k) \right) + A_{12}.$$

But, if $A_1^{(k)}$ is semi-positive definite, then $A_1^{(k)}(p_k - \underline{p}_k)$ is semi-positive definite as well; and if $A_2^{(k)}$ is semi-negative definite, then $A_2^{(k)}(p_k - \overline{p}_k)$ is semi-positive definite. So, $A(p)$ is positive semi-definite as a sum of positive semi-definite matrices. □

The splitting of $A^{(k)}$, as suggested in [84], can be done by using spectral decomposition. Since $A^{(k)}$ is symmetric, thus it is normal and according to the spectral theorem can be written as $A^{(k)} = QDQ^T$, where Q is a unitary matrix and D is a diagonal matrix with eigenvalues of $A^{(k)}$ on the diagonal. Since the eigenvalues of $A^{(k)}$ are real numbers, hence D can be split into $D = D^+ + D^-$, where $D^+ > 0$ and $D^- < 0$. Then, $A^{(k)} = A_1^{(k)} + A_2^{(k)}$, where $A_1^{(k)} = QD^+Q^T$ is positive definite and $A_2^{(k)} = Q + D^-Q^T$ is negative definite.

Theorems 3.51, 3.52, and 3.54 can be quite straightforwardly reformulated in terms of the positive definiteness [84]. The next theorem (cf. [84]) shows that there is a close relation between the positive definiteness and regularity of parametric interval matrices (cf. [212]).

Theorem 3.55 (Hladík [84]) *Parametric interval matrix $A(\boldsymbol{p})$ is positive definite if and only if the following two assertions hold:*

(i) $A(p)$ is positive definite for some $p \in \boldsymbol{p}$,
(ii) $A(\boldsymbol{p})$ is regular.

Proof The "only if" part is obvious, since each positive definite matrix is nonsingular. To prove the "if" part, suppose to the contrary that $A(\boldsymbol{p})$ is not positive definite. This and assumption (i) imply there are $A(p), A(p') \in A(\boldsymbol{p})$ such that $A(p)$ has all eigenvalues positive and $A(p')$ has at least one non-positive eigenvalue. But, since eigenvalues vary continuously with the entries of a matrix [141] and $\boldsymbol{p}$ is compact, hence there must be in $A(\boldsymbol{p})$ a matrix with at least one eigenvalue equal to zero. This means that $A(\boldsymbol{p})$ is singular which contradicts (ii). □

We will now give another sufficient verifiable condition for checking the positive definiteness of a parametric interval matrix (cf. [212]).

Theorem 3.56 *Let $A(\boldsymbol{p})$ be a parametric interval matrix such that $A(p^c)$ is positive definite and (3.45) holds true. Then, $A(\boldsymbol{p})$ is positive definite.*

Proof Condition (3.45) guarantees the regularity of $A(\boldsymbol{p})$. Since $A(p^c)$ is positive definite and $A(\boldsymbol{p})$ is regular, hence $A(\boldsymbol{p})$ is strongly positive definite (by Theorem 3.55). □

The above sufficient condition can be verified in polynomial time even with a large number of parameters. The spectral decomposition can be done in $\mathcal{O}(n^3)$ (for symmetric matrices, even faster using Wilkinson's algorithm, which has $\mathcal{O}(n^2)$ time complexity); the matrix inverse can also be done in $\mathcal{O}(n^3)$ (or even faster using the Coppersmith–Winograd algorithm, for example, which has $\mathcal{O}(n^{2.376})$ time complexity). Hence, we obtain asymptotic time complexity $\mathcal{O}(n^3)$.

Example 3.57 Hladík [84] considered cubic forms, since the entries of the Hessian of a cubic form are affine-linear functions of variables. However, using the affine transformation from Sect. 3.3, we can extend the applicability of the above theorems to a larger class of functions. Let us consider the following function of $x = (x_1, x_2, x_3)$ (cf. [84]):

$$f(x) = x_1^4 + 2x_1^2x_2 - x_1x_2x_3 + 3x_2x_3^2 + 5x_2^3$$

with $x \in \boldsymbol{x} = ([2, 3], [1, 2], [0, 1])$. Hessian matrix of f is given by

$$\nabla^2 f(x_1, x_2, x_3) = \begin{pmatrix} 12x_1^2 + 4x_2 & 4x_1 - x_3 & -x_2 \\ 4x_1 - x_3 & 30x_2 & 6x_3 - x_1 \\ -x_2 & 6x_3 - x_1 & 6x_2 \end{pmatrix}.$$

The affine transformation of $\nabla^2 f(x_1, x_2, x_3)$ yields the following parametric matrix with affine-linear dependencies:

$$A(e) = \begin{pmatrix} 82.5 + 30\varepsilon_1 + 2\varepsilon_2 + 1.5\varepsilon_4 & 9.5 + 2\varepsilon_1 - 0.5\varepsilon_3 & -1.5 - 0.5\varepsilon_2 \\ 9.5 + 2\varepsilon_1 - 0.5\varepsilon_3 & 45 + 15\varepsilon_2 & 0.5 - 0.5\varepsilon_1 + 3\varepsilon_3 \\ -1.5 - 0.5\varepsilon_2 & 0.5 - 0.5\varepsilon_1 + 3\varepsilon_3 & 9 + 3\varepsilon_2 \end{pmatrix}.$$

Since $A(p^c) = A^{(0)}$ is positive definite with eigenvalues of $(8.96, 42.75, 84.79)$ and $\rho(|(A^{(0)})^{-1}|B) \approx 0.61 < 1$, (rounded to two decimal places), $\nabla^2 f(\boldsymbol{x})$ is positive definite by Theorem 3.56; thus, f is strictly convex on $\boldsymbol{x}$.

3.10 Hurwitz Stability

A real square matrix A is called *stable matrix* (also *Hurwitz matrix*) if all of its eigenvalues have a strictly negative real part, i.e., if, for each eigenvalue λ_i of A, $\mathrm{Re}(\lambda_i) < 0$. The problem of the stability of interval matrices is strongly connected with the behavior of a linear time-invariant system $\dot{x}(t) = Ax(t)$ under perturbation (see, e.g., [13, 132, 134], and the references therein). As in [212], we mainly investigate the stability of symmetric parametric interval matrices with affine-linear dependencies here. The great advantage of the parametric approach is that a parametric interval matrix contains only symmetric matrices, whereas an interval matrix can also contain non-symmetric matrices (whose eigenvalues might not be real); therefore, it requires some additional care [212].

Definition 3.58 Parametric interval matrix $A(\boldsymbol{p})$ is stable if, for each $p \in \boldsymbol{p}$, $A(p)$ is stable.

Lemma 3.59 (cf. Fiedler [51]) *Real symmetric matrix A is stable if and only if $-A$ is positive definite.*

In the first theorem, we give some sufficient and necessary conditions for the stability of symmetric parametric interval matrices (cf. [212]).

Theorem 3.60 *Let $A(\boldsymbol{p})$ be a symmetric parametric interval matrix. Then, the following assertions are equivalent:*

(i) $A(\boldsymbol{p})$ is stable,
(ii) $A(p)$ is stable for each p such that $p_k \in \{\underline{p}_k, \overline{p}_k\}$ (vertex property),
(iii) $-A(\boldsymbol{p})$ is strongly positive definite,

Proof Condition (i) is equivalent to (iii) by Lemma 3.59. So, it is enough to prove (i)⇒(ii)⇒(iii).

(i)⇒(ii) The proof is obvious since, for $A(p)$ such that $p_k \in \{\underline{p}_k, \overline{p}_k\}$, it holds that $A(p) \in A(\boldsymbol{p})$.

(ii)⇒(iii) Let p be such that $p_k \in \{\underline{p}_k, \overline{p}_k\}$. Then, $A(p)$ is stable; thus, by Lemma 3.59, $-A(p)$ is positive definite. Since p was chosen arbitrarily, $-A(\boldsymbol{p})$ is strongly positive definite (cf. Theorem 3.51). □

In the next theorem, we formulate another sufficient and necessary condition that provides a link between the stability and regularity of a symmetric parametric interval matrix (cf. [212]).

Theorem 3.61 *A symmetric parametric interval matrix $A(\boldsymbol{p})$ is stable if and only if the following two assertions hold:*

(i) $A(p)$ is stable for an arbitrary $p \in \boldsymbol{p}$,
(ii) $A(\boldsymbol{p})$ is regular.

Proof The "only if" part is obvious, since each stable matrix is non-singular. Conversely, if $A(p)$ is stable, then $-A(p)$ is positive definite; and since $A(\boldsymbol{p})$ is regular, $-A(\boldsymbol{p})$ is regular as well. Thus, by Theorem 3.55, $-A(\boldsymbol{p})$ is strongly positive definite, and $A(\boldsymbol{p})$ is therefore stable by Theorem 3.60.

□

Theorem 3.62 *Let $A(\boldsymbol{p})$ be a symmetric parametric interval matrix such that $A(p^c)$ is stable and (3.45) holds true. Then, $A(\boldsymbol{p})$ is stable.*

Proof The theorem follows directly from Theorem 3.56 applied to $-A(\boldsymbol{p})$. □

3.11 Schur Stability

A real square matrix A is called *Schur stable* if all of its eigenvalues lie in a unit circle, i.e., if $|\lambda_i| < 1$ for each eigenvalue λ_i of A. Schur matrices are strongly connected with the asymptotic stability of polynomials and dynamical systems. Similar to the previous section, we only consider symmetric parametric interval matrices (cf. [212]) with affine-linear dependencies here.

Definition 3.63 A parametric interval matrix $A(\boldsymbol{p})$ is Schur stable if, for each $p \in \boldsymbol{p}$, $A(p)$ is Schur stable.

The first theorem we present gives a necessary and sufficient condition for a symmetric parametric interval matrix to be Schur stable. A similar result for interval matrices was given in [77].

Theorem 3.64 *Let $A(\boldsymbol{p})$ be a symmetric parametric interval matrix. Then, the following assertions are equivalent:*

(i) $A(\boldsymbol{p})$ is Schur stable,
(ii) $A(p)$ is Schur stable for each p such that $p_k \in \{\underline{p}_k, \overline{p}_k\}$.

Proof Implication (i)⇒(ii) is obvious. To prove (ii)⇒(i), take arbitrary $A(p') \in A(\boldsymbol{p})$, $x \neq 0$ and put $s_k = \operatorname{sgn}(x^T A^{(k)} x)$. Since

$$\left|x^T(A(p') - A(p^c))x\right| \leqslant \sum_{k=1}^{K} \left|x^T A^{(k)} x\right| p_k^{\Delta},$$

$$\begin{aligned} x^T A(p')x &= x^T A(p^c)x + x^T(A(p') - A(p^c))x \leqslant \\ &\leqslant x^T A(p^c)x + \sum_{k=1}^{K} \left|x^T A^{(k)} x\right| p_k^{\Delta} = \\ &= x^T \left(A(p^c) + \sum_{k=1}^{K} A^{(k)} s_k p_k^{\Delta} \right) x, \end{aligned}$$

and we have

$$\frac{x^T A(p')x}{x^T x} \leqslant \frac{x^T \left(A(p^c) + \sum_{k=1}^{K} A^{(k)} s_k p_k^{\Delta}\right) x}{x^T x} = \frac{x^T A(p)x}{x^T x},$$

where p is such that $p_k \in \{\underline{p}_k, \overline{p}_k\}$, which implies that

$$\lambda_{\max}(A(p')) \leqslant \lambda_{\max}(A(p)) < 1.$$

Analogically, we can prove that $\lambda_{\min}(A(p')) > -1$; therefore, $A(p')$ is Schur stable. Since $A(p')$ was chosen arbitrarily, $A(\boldsymbol{p})$ is Schur stable. □

The next theorem gives another sufficient and necessary condition for the Schur stability of symmetric parametric interval matrices. It also links Schur stability and Hurwitz stability (cf. [212]).

Theorem 3.65 *Let $A(\boldsymbol{p})$ be a symmetric parametric interval matrix. Then, $A(\boldsymbol{p})$ is Schur stable if and only if parametric interval matrices*

$$A(\boldsymbol{p}) - I \quad \textit{and} \quad -A(\boldsymbol{p}) - I$$

are stable.

Proof To prove the "only if" part, take arbitrary $p \in \boldsymbol{p}$. Since $A(p)$ is Schur stable, hence all its eigenvalues are within the disk $(-1, 1)$. Therefore, all eigenvalues of $A(p) - I$ lie within $(-2, 0)$, and $A(\boldsymbol{p}) - I$ is therefore stable. To prove the stability of matrix $-A(\boldsymbol{p}) - I$, it is enough to observe that, if $A(p)$ is Schur stable, then $-A(p)$ is Schur stable as well.

To prove the "if" part, take $A(p) \in A(\boldsymbol{p})$ and let λ be its eigenvalue. Since $A(p) - I$ is stable, all of its eigenvalues are negative; hence, $\lambda - 1 < 0$, which means that $\lambda < 1$. Moreover, $-\lambda$ is an eigenvalue of $-A(p)$, and since $-A(p) - I$ is stable, we obtain that $\lambda > -1$ using the same reasoning. Since $A(p)$ was chosen arbitrarily, $A(\boldsymbol{p})$ is Schur stable. □

In a similar manner as in the case of Hurwitz stability, we can formulate the sufficient condition for the Schur stability of symmetric parametric interval matrices. The validity of the theorem follows directly from Theorems 3.65 and 3.62.

Theorem 3.66 *Let $A(\boldsymbol{p})$ be a symmetric parametric interval matrix such that $A(p^c)$ is Schur stable and conditions*

$$\rho\left(\left|\left(A\left(p^c\right) - I\right)^{-1} B\right|\right) < 1 \tag{3.72a}$$

$$\rho\left(\left|\left(-A\left(p^c\right) - I\right)^{-1} B\right|\right) < 1 \tag{3.72b}$$

are satisfied with B given by formula (3.44). Then, $A(\boldsymbol{p})$ is Schur stable.

3.12 Radius of Stability

The radius of stability $s(A(\boldsymbol{p}))$ of a parametric interval matrix $A(\boldsymbol{p})$ is given by formula (3.42). Obviously, if $s(A(\boldsymbol{p})) = 0$, then $A(\boldsymbol{p})$ is unstable, and if $s(A(\boldsymbol{p})) > 1$, then $A(\boldsymbol{p})$ is stable.

Proposition 3.67 *Let $A(\boldsymbol{p})$ be a symmetric parametric interval matrix such that $A(p^c)$ is stable. Then,*

$$s(A(\boldsymbol{p})) = r(A(\boldsymbol{p})) = \frac{1}{\rho_0(A(\boldsymbol{p}))}, \tag{3.73}$$

where $\rho_0(A(\boldsymbol{p}))$ is given by formula (3.65).

Proof The proposition follows directly from Theorems 3.61 and 3.47. □

3.13 Inverse Stability

Definition 3.68 A regular parametric interval matrix $A\boldsymbol{p})$ is *inverse (sign) stable* if, for each $i, j = 1, \ldots, n$, either $A(p)^{-1}_{ij} \leqslant 0$ for each $p \in \boldsymbol{p}$ or $A(p)^{-1}_{ij} \geqslant 0$ for each $p \in \boldsymbol{p}$.

Given parametric interval matrix $A(\boldsymbol{p})$, let

$$D = \left|A(p^c)^{-1}\right| B, \tag{3.74}$$

where B is given by formula (3.44), and let

$$G = D(I - D)^{-1}. \tag{3.75}$$

The following theorem (cf. Theorem 4.4 from [208]) gives a verifiable sufficient criterion of inverse stability of a parametric interval matrix.

Theorem 3.69 *Let $A(\boldsymbol{p})$ be a square parametric interval matrix such that $A(p^c)$ is non-singular and (3.45) is satisfied. If*

$$G|A(p^c)^{-1}| \leqslant |A(p^c)^{-1}|,$$

where G is given by formula (3.75), then $A(\boldsymbol{p})$ is inverse stable.

Proof Let $p \in \boldsymbol{p}$. Then,

$$A(p) = A(p^c)\left(I - A(p^c)^{-1}(A(p^c) - A(p))\right) \tag{3.76}$$

and

$$\rho\left(A\left(p^c\right)^{-1}\left(A(p^c)-A(p)\right)\right)\leqslant\rho\left(\left|A(p^c)^{-1}\right|B\right)<1. \tag{3.77}$$

From (3.76) and (3.77) it follows that

$$A(p)^{-1}=\left(\sum_{j=0}^{\infty}\left(A(p^c)^{-1}(A(p^c)-A(p))\right)^j\right)A\left(p^c\right)^{-1}.$$

So, we have

$$\begin{aligned}\left|A(p)^{-1}-A(p^c)^{-1}\right| &= \left|\sum_{j=1}^{\infty}\left(A(p^c)^{-1}(A(p^c)-A(p))\right)^j\right|\left|A(p^c)^{-1}\right|\leqslant\\ &\leqslant\sum_{j=1}^{\infty}\left|A(p^c)^{-1}(A(p^c)-A(p))\right|^j\left|A(p^c)^{-1}\right|\leqslant\\ &\leqslant\sum_{j=1}^{\infty}D^j\left|A(p^c)^{-1}\right|=G\left|A(p^c)^{-1}\right|\leqslant\left|A(p^c)^{-1}\right|.\end{aligned}$$

Hence, if $A(p^c)^{-1}_{ij}\geqslant 0$, then $A(p)^{-1}_{ij}\geqslant 0$, and if $A(p^c)^{-1}_{ij}\leqslant 0$, then $A(p)^{-1}_{ij}\leqslant 0$ (which proves that $A(\boldsymbol{p})$ is inverse stable). □

The next theorem (cf. [210]) gives another verifiable sufficient condition for inverse stability. It uses matrix R specified only by some inequality, which enables us to use in practical computation the numerically computed inverse the midpoint matrix.

Theorem 3.70 *Let $A(\boldsymbol{p})$ be a parametric interval matrix. Let B be given by formula (3.44) and let R be such that matrix*

$$G_R=\left|I-RA(p^c)\right|+|R|B$$

satisfies

$$2G_R|R|<|R|. \tag{3.78}$$

Then $A(\boldsymbol{p})$ is inverse stable and the sigh pattern of each inverse matrix is identical with that of R. Moreover, R is non-singular and $\rho(G_R)<\frac{1}{2}$.

Proof Put $r=|R_{*i}|$, where R_{*i} denotes the ith row of R. Since by (3.78) $r>0$ and $2G_Rr<r$, thus (cf. Corollary 3.4) $\rho(G_R)<\frac{1}{2}$. Moreover, $(I-G_R)^{-1}$ exists and is non-negative. Now, for each $A(p)\in A(\boldsymbol{p})$, we have

$$RA(p)=I-(I-R(A(p)). \tag{3.79}$$

Since

$$|I - RA(p)| = |I - RA(p^c) + R(A(p^c) - A(p))| \leqslant |I - RA(p^c)| + |R|B,$$

hence $\rho(I - RA(p)| < \frac{1}{2}$. This means that $RA(p)$ is non-singular, which implies that R is non-singular and $A(\boldsymbol{p})$ is regular; and from (3.79) it follows that

$$A(p)^{-1} = \left(\sum_{j=0}^{\infty} (I - RA(p))^j \right) R.$$

Using similar reasoning as in the proof of Theorem 3.69, we obtain

$$|A(p)^{-1} - R| < |R|. \tag{3.80}$$

Thus, id $R_{ij} > 0$, then $A(p)_{ij} > 0$ and if $R_{ij} < 0$, then $A(p)_{ij} < 0$ (on account of (3.78) the case $R_{ij} = 0$ cannot occur), which proves that $A(\boldsymbol{p})$ is inverse stable. □

Chapter 4
Linear Systems

In this chapter we define *parametric interval linear systems* and characterize various types of parametric solution sets. We provided as well several visualizations of the most general (united) solution set (in 2D and 3D) that aim to facilitate the exploration of the properties of this solution set. We start with a brief introduction to *interval linear systems* [160, 206] from which parametric interval linear systems originate.

4.1 Interval Linear Systems

An interval linear system with coefficient matrix $\boldsymbol{A} \in \mathbb{IR}^{n\times n}$ and right-hand vector $\boldsymbol{b} \in \mathbb{IR}^n$ is defined as family of linear equations

$$Ax = b,\ \ A \in \boldsymbol{A}, b \in \boldsymbol{b}, \tag{4.1}$$

and is usually written shortly as

$$\boldsymbol{A}x = \boldsymbol{b}. \tag{4.2}$$

Generalized solution sets [244] of system (4.2) are defined by

$$\left\{x \in \mathbb{R}^n \middle| \left(Q_1 C_{\pi(1)} \in \boldsymbol{C}_{\pi(1)}\right) \ldots \left(Q_m C_{\pi(m)} \in \boldsymbol{C}_{\pi(m)}\right)\ :\ Ax = b\right\} \tag{4.3}$$

where $m = n^2 + n$, $Q_i \in \{\forall, \exists\}$ $(i = 1, \ldots, m)$ are logical quantifiers, $C = (A_{11}, \ldots, A_{nn}, b_1, \ldots, b_n)$ is the aggregated vector of the entries of A and b, $\boldsymbol{C} = (\boldsymbol{A}_{11}, \ldots, \boldsymbol{A}_{nn}, \boldsymbol{b}_1, \ldots, \boldsymbol{b}_n)$ is the aggregated vector of the intervals of the possible values of the entries of A and b, and $\pi : \mathbb{R}^m \to \mathbb{R}^m$ is a permutation. Since the total number of solution sets that can be defined for system (4.2) by using formula (4.3) exceeds 2^m, only those generalized solution sets are usually treated in which all occurrences of the universal quantifiers precede all occurrences of the existential

I. Skalna, *Parametric Interval Algebraic Systems*, Studies in Computational Intelligence 766, https://doi.org/10.1007/978-3-319-75187-0_4

quantifiers. Because of this constraint, they are called the *AE-solution sets*. A useful representation of AE-solution sets was given in [244]:

$$S_{AE}(\boldsymbol{A}, \boldsymbol{b}) = \Big\{ x \in \mathbb{R}^n \mid \left(\forall A^{\forall} \in \boldsymbol{A}^{\forall}\right) \left(\forall b^{\forall} \in \boldsymbol{b}^{\forall}\right) \left(\exists A^{\exists} \in \boldsymbol{A}^{\exists}\right) \left(\exists b^{\exists} \in \boldsymbol{b}^{\exists}\right) : \left(A^{\forall} + A^{\exists}\right) x = \left(b^{\forall} + b^{\exists}\right) \Big\}, \quad (4.4)$$

where, for α, Q_{ij}, $Q_i \in \{\forall, \exists\}$,

$$\boldsymbol{A}_{ij}^{\alpha} = \begin{cases} \boldsymbol{A}_{ij}, & Q_{ij} = \alpha \\ 0, & \text{otherwise} \end{cases}, \quad \boldsymbol{b}_i^{\alpha} = \begin{cases} \boldsymbol{b}_i, & Q_i = \alpha \\ 0, & \text{otherwise} \end{cases}. \quad (4.5)$$

The (united) solution set of system (4.2), formed by the solutions to all point systems from family (4.1), is defined by

$$S(\boldsymbol{A}, \boldsymbol{b}) \triangleq \{x \in \mathbb{R}^n \mid (\exists A \in \boldsymbol{A}) \; (\exists b \in \boldsymbol{b}) \; : \; Ax = b\}. \quad (4.6)$$

It is undoubtedly the most popular AE-solution set, therefore it is often referred to simply as solution set of an interval linear system. The wide popularity of the united solution set is probably due to the close relationship of interval linear systems with sensitivity problems.

The following theorem gives a description of $S(\boldsymbol{A}, \boldsymbol{b})$ by means of a set of inequalities.

Theorem 4.1 (Oettli and Prager [166]) *Let* $\boldsymbol{A} \in \mathbb{IR}^{n \times n}$, $\boldsymbol{b} \in \mathbb{IR}^n$. *Then,*

$$S(\boldsymbol{A}, \boldsymbol{b}) = \{x \in \mathbb{R}^n \mid |A^c x - b^c| \leqslant A^{\Delta}|x| + b^{\Delta}\}. \quad (4.7)$$

It can be seen from this description that the intersection of $S(\boldsymbol{A}, \boldsymbol{b})$ with each orthant is a convex polyhedron [16]. Hence, $S(\boldsymbol{A}, \boldsymbol{b})$ as a sum of disjoint convex polyhedra is generally non-convex.

Two other particular cases of AE-solution sets that are subject to research in interval analysis are:

- *tolerable solution set*

$$\begin{aligned} S_{tol}(\boldsymbol{A}, \boldsymbol{b}) &= \{x \in \mathbb{R}^n \mid (\forall A \in \boldsymbol{A}) \; (\exists b \in \boldsymbol{b}) \; : \; Ax = b\} = \\ &= \{x \in \mathbb{R}^n \mid \boldsymbol{A}x \subseteq \boldsymbol{b}\}, \end{aligned} \quad (4.8)$$

  formed by all point vectors $x \in \mathbb{R}^n$ such that the image $Ax \in \boldsymbol{b}$ for all $A \in \boldsymbol{A}$ (see, e.g., [158, 160, 242, 243]).

– *controllable solution set*

$$S_{ctr}(\boldsymbol{A}, \boldsymbol{b}) = \{x \in \mathbb{R}^n \mid (\forall b \in \boldsymbol{b})\ (\exists A \in \boldsymbol{A})\ :\ Ax = b\} = \\ = \{x \in \mathbb{R}^n \mid \boldsymbol{A}x \supseteq \boldsymbol{b}\}, \tag{4.9}$$

formed by all point vectors $x \in \mathbb{R}^n$ such that for selected $b \in \boldsymbol{b}$ there is $A \in \boldsymbol{A}$ satisfying $Ax = b$ (see, e.g., [239]).

4.2 Parametric Interval Linear Systems

A parametric interval linear system is given by

$$A(\boldsymbol{p})x = b(\boldsymbol{p}), \tag{4.10}$$

where $\boldsymbol{p} \in \mathbb{R}^K$ is a vector of interval parameters, $A(\boldsymbol{p})$ is a parametric interval matrix, and $b(\boldsymbol{p})$ is a parametric interval vector. According to the definition of a parametric interval matrix, this means that a parametric interval linear system is, in fact, the following family of real parametric linear systems:

$$\{A(p)x = b(p),\ p \in \boldsymbol{p}\}. \tag{4.11}$$

By analogy to the classical interval case, generalized solution sets are defined for parametric system (4.11) as follows:

Definition 4.2 Generalized parametric solution sets of system (4.10) are defined by

$$\left\{x \in \mathbb{R}^n \mid \left(Q_1\, p_1 \in \boldsymbol{p}_1\right) \ldots \left(Q_K\, p_K \in \boldsymbol{p}_K\right)\ :\ A(p)x = b(p)\right\}, \tag{4.12}$$

where $Q_i \in \{\forall, \exists\}$, $i = 1, \ldots, K$.

According to the above definition, different quantification of parameters in their interval domains is allowed. However, similarly as in the interval case, only parametric AE-solution sets are in practice considered (cf. [180]):

$$S_{AE}(\boldsymbol{p}) = \left\{x \in \mathbb{R}^n \mid \left(\forall p^{\forall} \in \boldsymbol{p}^{\forall}\right) \left(\exists p^{\exists} \in \boldsymbol{p}^{\exists}\right)\ : A(p)x = b(p)\ \right\}, \tag{4.13}$$

where $\left(p^{\forall}, p^{\exists}\right) = \left(p_{\pi(1)}, \ldots, p_{\pi(K)}\right)$, $\left(\boldsymbol{p}^{\forall}, \boldsymbol{p}^{\exists}\right) = \left(\boldsymbol{p}_{\pi(1)}, \ldots, \boldsymbol{p}_{\pi(K)}\right)$, and π is a permutation.

The parametric (united) solution set of parametric interval linear system (4.10), formed by solutions of all point systems from family (4.11), is of primary interest in this book.

Definition 4.3 The parametric (united) solution set of system (4.10) is defined by

$$S(\boldsymbol{p}) \triangleq \{x \in \mathbb{R}^n \mid \exists p \in \boldsymbol{p} \ : \ A(p)x = b(p)\}. \tag{4.14}$$

Proposition 4.4 *If $A(\boldsymbol{p})$ is regular, then solution set $S(\boldsymbol{p})$ is compact and connected.*

If $A(\boldsymbol{p})$ is singular, then the solution set is either empty (which is rare) or unbounded.

The other two parametric AE-solution sets that are studied in the literature, i.e., the *parametric tolerable solution set* and the *parametric controllable solution*, can be defined as follows (cf. [181, 186]).

Definition 4.5 The parametric tolerable solution set of system (4.10) is defined by

$$S_{tol}(\boldsymbol{p}) \triangleq \{x \in \mathbb{R}^n \mid (\forall p^{\forall} \in \boldsymbol{p}^{\forall}) \ (\exists p^{\exists} \in \boldsymbol{p}^{\exists}) \ : \ A(p^{\forall})\, x = b(p^{\exists})\}, \tag{4.15}$$

where $(p^{\forall}, p^{\exists}) = (p_{\pi(1)}, \ldots, p_{\pi(K)})$, $(\boldsymbol{p}^{\forall}, \boldsymbol{p}^{\exists}) = (\boldsymbol{p}_{\pi(1)}, \ldots, \boldsymbol{p}_{\pi(K)})$, and π is a permutation.

Definition 4.6 The parametric controllable solution set of the system (4.10) is defined by

$$S_{ctr}(\boldsymbol{p}) \triangleq \{x \in \mathbb{R}^n \mid (\forall p^{\forall} \in \boldsymbol{p}^{\forall}) \ (\exists p^{\exists} \in \boldsymbol{p}^{\exists}) \ : \ A(p^{\exists})\, x = b(p^{\forall})\}, \tag{4.16}$$

where $(p^{\forall}, p^{\exists}) = (p_{\pi(1)}, \ldots, p_{\pi(K)})$, $(\boldsymbol{p}^{\forall}, \boldsymbol{p}^{\exists}) = (\boldsymbol{p}_{\pi(1)}, \ldots, \boldsymbol{p}_{\pi(K)})$, and π is a permutation.

AE-solution sets of classical interval linear systems have been extensively studied, e.g., in [64, 65, 239, 241–244, 272] (see also the references given therein). However, there are only a few results on general case of AE-solution sets of parametric interval linear systems. Most of them have appeared quite recently and are concerned mainly with the solution sets of parametric interval linear systems with affine-linear dependencies. A special case of parametric tolerable solution sets was considered in [240], and the unbounded tolerable solution set was investigated in [183]. AE-solution sets with linear shape were discussed, e.g., in [85, 184], whereas AE solution sets of a class of parametric interval linear systems were characterized in [191]. An explicit description of parametric AE-solution sets was given in [180], and outer interval enclosures for the parametric AE solution sets were discussed in [186].

4.3 Shape of (United) Parametric Solution Set

The structure of the (united) parametric solution set is usually quite complex. It is more complex than the solution set of classical interval linear systems, since it is generally not convex (even in a single orthant). In particular, the solution set of linear systems with symmetric matrices (representing quite-trivial dependencies) is a sum of compact sets, the boundaries of which are quadrics [6]. An explicit, sufficient, and necessary characterization of the solution set of systems with symmetric and skew-symmetric matrices by means of nonlinear inequalities was given in [78]. The solution sets of systems with symmetric and skew-symmetric matrices were also investigated in [137, 138].

If there are only linear dependencies in (4.10), the united solution set (4.14) can be described by the following set of trivial inequalities [180]:

$$\left(A_{ij}^{(0)} + \sum_{k=1}^{K} A_{ij}^{(k)} \underline{p}_k\right) x_i \leqslant b_i^{(0)} + \sum_{k=1}^{K} b_i^{(k)} p_k \leqslant \tag{4.17a}$$

$$\leqslant \left(A_{ij}^{(0)} + \sum_{k=1}^{K} A_{ij}^{(k)} p_k\right) x_i,\ i = 1, \ldots, n,$$

$$\underline{p}_k \leqslant p_k \leqslant \overline{p}_k,\ k = 1, \ldots, K. \tag{4.17b}$$

In order to obtain the inequalities not involving parameters p_k, we can employ the Fourier–Motzkin elimination (see, e.g., [237]). The next theorem shows one step of the elimination process.

Theorem 4.7 (Alefeld et al. [7]) *Let $f_{\lambda\mu}$, g_λ, $\lambda = 1, \ldots, k$ ($k \geqslant 2$), $\mu = 1, \ldots, m$, be real-valued functions of $x = (x_1, \ldots, x_n)^T$ on some subset $D \subseteq \mathbb{R}^n$. Assume that there are non-negative integers $k_1 < k_2 < k$ such that $f_{\lambda 1}(x) \not\equiv 0$ for all $\lambda \in \{k_1 + 1, \ldots, k\}$ and $f_{\lambda 1}(x) \geqslant 0$ for all $x \in D$ and all $\lambda \in \{k_1 + 1, \ldots, k\}$; also, for each $x \in D$, there is an index $\beta^* \in \{k_1 + 1, \ldots, k_2\}$ with $f_{\beta^* 1}(x) > 0$ and an index $\gamma^* \in \{k_2 + 1, \ldots, k\}$ with $f_{\gamma^* 1}(x) > 0$. For m parameters, $p_1, \ldots, p_m$. For m parameters, $p_1, \ldots, p_m$ varying in $\mathbb{R}$, and for x varying in D, define the two sets of inequalities (4.18) and (4.19) (in both sets, trivial inequalities such as $0 \leqslant 0$ are omitted).*

$$g_\alpha(x) + \sum_{\mu=2}^{m} f_{\alpha\mu}(x) p_\mu \leqslant 0,\ \alpha = 1, \ldots, k_1, \tag{4.18a}$$

$$g_\beta(x) + \sum_{\mu=2}^{m} f_{\beta\mu}(x) p_\mu \leqslant f_{\beta 1}(x) p_1,\ \beta = k_1 + 1, \ldots, k_2, \tag{4.18b}$$

$$f_{\gamma 1}(x) p_1 \leqslant g_\gamma + \sum_{\mu=2}^{m} f_{\gamma\mu} p_\mu,\ \gamma = k_2 + 1, \ldots, k. \tag{4.18c}$$

$$g_\alpha(x) + \sum_{\mu=2}^{m} f_{\alpha\mu}(x)p_\mu \leqslant 0, \ \alpha = 1, \ldots, k_1, \tag{4.19a}$$

$$g_\beta(x)f_{\gamma 1}(x) + \sum_{\mu=2}^{m} f_{\beta\mu}(x)f_{\gamma 1}(x)p_\mu \leqslant$$

$$\leqslant g_\gamma(x)f_{\beta 1}(x) + \sum_{\mu=2}^{m} f_{\gamma\mu}(x)f_{\beta 1}(x)p_\mu, \tag{4.19b}$$

$$\beta = k_1 + 1, \ldots, k_2, \ \gamma = k_2 + 1, \ldots, k. \tag{4.19c}$$

Then, for any set $T \subseteq \mathbb{R}^n$, the following assertions are equivalent:

(a) $T \cap D$ is described by set of inequalities (4.18),
(b) $T \cap D$ is described by set of inequalities (4.19).

Since p_1 no longer occurs in (4.19), we will call the transition from (4.18) to (4.19) the elimination of p_1 from (4.18).

As pointed out in [179], the elimination process described above is lengthy and not at all unique. Moreover, the number of final inequalities grows exponentially with the number of parameters and the size of a problem [6, 179]. So, this approach is generally not recommended for practical use. However, if a parametric interval linear system involves only *zero* and *first class* parameters, then its solution set is characterized by $2n$ linear inequalities [180]. A parameter is of class zero if it is involved in the right-hand side of only one equation of a parametric interval linear system, whereas a first class parameter is the one that is not of class zero and occurs in only one equation of a parametric interval linear system, however there is no limit on the number of occurrences.

Theorem 4.8 (Popova [180]) *Let $A(\boldsymbol{p})x = b(\boldsymbol{p})$ be a parametric interval linear system with affine-linear dependencies involving only parameters of zero or first class. Then, solution set $S(\boldsymbol{p})$ is characterized by*

$$S(\boldsymbol{p}) = \left\{ x \in \mathbb{R}^n \;\middle|\; \left|A(p^c)x - b(p^c)\right| \leqslant \sum_{k=1}^{K} \left|b^{(k)} - A^{(k)}\right| p_k^{\Delta} \right\}. \tag{4.20}$$

The set of inequalities obtained from the Fourier–Motzkin elimination can be used to draw the parametric solution set in environments supporting tools for inequality plotting [180].

The characterization of the boundary of parametric solution set $S(\boldsymbol{p})$ by means of pieces of *parametric hypersurfaces* (PHS) was proposed in [190]. The general considerations presented there concerned mainly parametric interval linear systems with affine-linear dependencies.

Definition 4.9 A parametric hypersurface in an n-dimensional space is defined by n parametric functions

$$x_i = x_i(p_1, \ldots, p_m),\ i = 1, \ldots, n,$$

where $m = n - 1$.

Parametric function x_i are defined by the analytic solution of parametric linear system $A(p)x = b(p)$. For example, for $n = 2, m = 1$, the parametric solution set is a curve defined by two parametric functions:

$$x_1(p) = \left\{A(p)^{-1}b(p)\right\}_1,$$
$$x_2(p) = \left\{A(p)^{-1}b(p)\right\}_2.$$

The approach based on hypersurfaces can be easily implemented in environments supporting functions for plotting parametric curves and surfaces [190] such as Mathematica or Maple.

Below, we present a visualization of the parametric solution sets of PILS with arbitrary dependencies, as such systems are of our main concern. The graphs visualizing the parametric solution sets in the following examples were generated using a simple Matlab code.

Example 4.10 Consider the following two-dimensional parametric interval linear system with affine-linear dependencies:

$$\begin{pmatrix} p_1 & 1 + p_2 \\ -2 & 3p_1 - 1 \end{pmatrix} \begin{pmatrix} x_1 \\ x_2 \end{pmatrix} = \begin{pmatrix} 2p_1 \\ 0 \end{pmatrix} \tag{4.21}$$

with $p_1, p_2 \in [0, 1]$. For each $p \in \boldsymbol{p}$, the determinant of $A(p)$ is a continuous function of p, which can be represented by the following expression: $\det(A(p)) = 3(p_1 - 1/6)^2 + 2p_2 + 23/12$. Since each parameter p_i occurs only once in the last expression, $\mathrm{Rge}\,(\det(A(p)) \mid \boldsymbol{p}) = [23/12, 6]$. This means that $A(\boldsymbol{p})$ is regular, thus, the solution set is compact and connected. Now, if we neglect the dependencies in (4.21) (we substitute $A(p)_{ij}, b_i, i, j = 1, \ldots, n$, by their natural interval extensions), then we will obtain the following interval linear system:

$$\begin{pmatrix} [0, 1] & [1, 2] \\ -2 & [-1, 2] \end{pmatrix} \begin{pmatrix} x_1 \\ x_2 \end{pmatrix} = \begin{pmatrix} [0, 2] \\ 0 \end{pmatrix}. \tag{4.22}$$

Solution set $S(\boldsymbol{A}, \boldsymbol{b})$ of system (4.22) can be determined from Theorem 4.1. Figure 4.1 shows the solution set of system (4.22) and the parametric solution set of system (4.21), the latter being represented by a set of plane parametric curves. Each parametric curve is defined using two functions $x_1(p_i)$ and $x_2(p_i)$, where p_i is one of the two parameters. By sampling the second parameter with step size δ, we obtain a set of parametric curves. Functions x_1 and x_2 are obtained from an analytical solution of system (4.21).

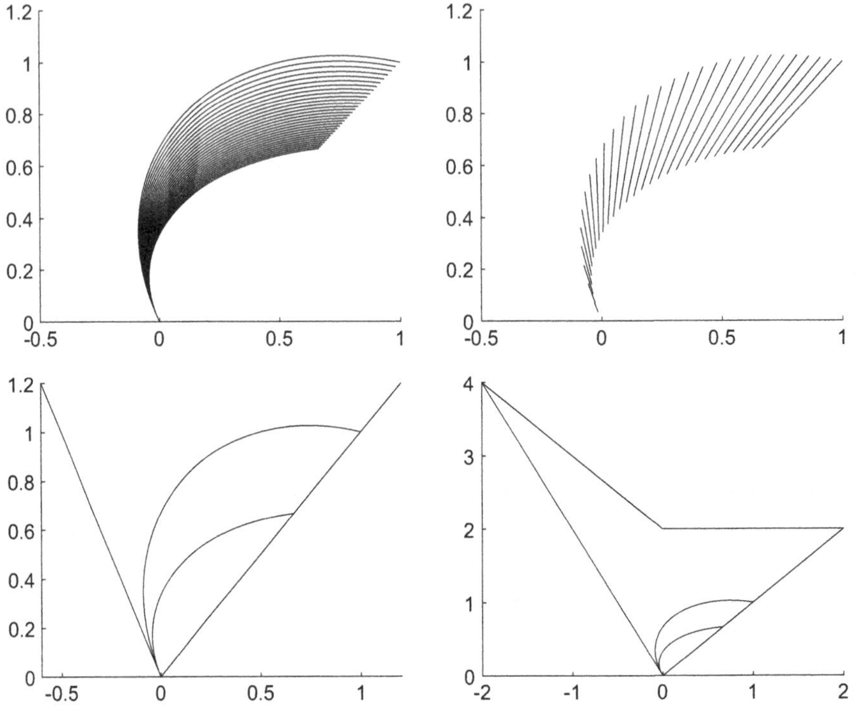

Fig. 4.1 Parametric solution set of system (4.21) depicted as set of plane curves depending on parameter p_1, $\delta \approx 0.03$ (top left) and as set of plane curves depending on parameter p_2, $\delta \approx 0.03$ (top right); parametric solution set with part of corresponding non-parametric solution set (bottom left) and with entire non-parametric solution set (bottom right)

Example 4.11 Consider the following two-dimensional parametric interval linear system with nonlinear dependencies:

$$\begin{pmatrix} p_1^4 & p_2 - 1 \\ p_2 & p_1 \end{pmatrix} \begin{pmatrix} x_1 \\ x_2 \end{pmatrix} = \begin{pmatrix} -p_2 + 1/3 \\ p_1^2 p_2 \end{pmatrix} \tag{4.23}$$

with parameters $p_1 \in [-2, -1]$, $p_2 \in [3, 5]$. It is not hard to verify that $\mathrm{Rge}\,(\det(A(p)) \mid \boldsymbol{p}) = [-52, -7]$, which means that $A(\boldsymbol{p})$ is regular; thus, the solution set is compact and connected. Figure 4.2 depicts the parametric solution set of system (4.23) as a set of plane parametric curves.

Example 4.12 Consider the following three-dimensional parametric interval linear system with nonlinear dependencies in the right-hand vector:

$$\begin{pmatrix} 1 & p_2^2 & p_2 \\ p_1 & 2 & p_1 \\ p_2 & p_1 & 3 \end{pmatrix} \begin{pmatrix} x_1 \\ x_2 \\ x_3 \end{pmatrix} = \begin{pmatrix} 1 \\ p_1^2 \\ p_2^2 \end{pmatrix}, \tag{4.24}$$

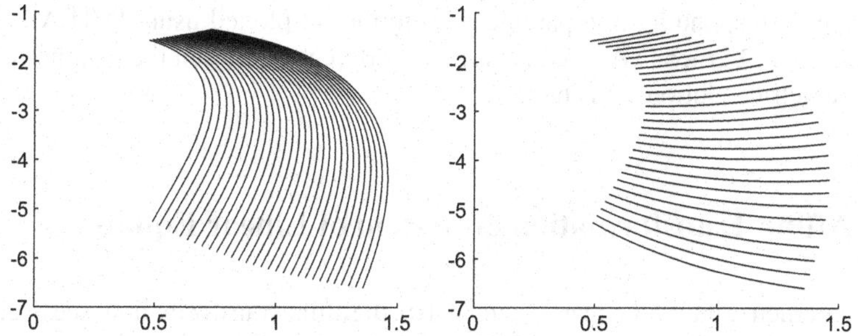

Fig. 4.2 Parametric solution set of system (4.23) depicted as set of plane curves depending on parameter p_1, $\delta \approx 0.03$ (left) and as set of plane curves depending on parameter p_2, $\delta \approx 0.07$ (right)

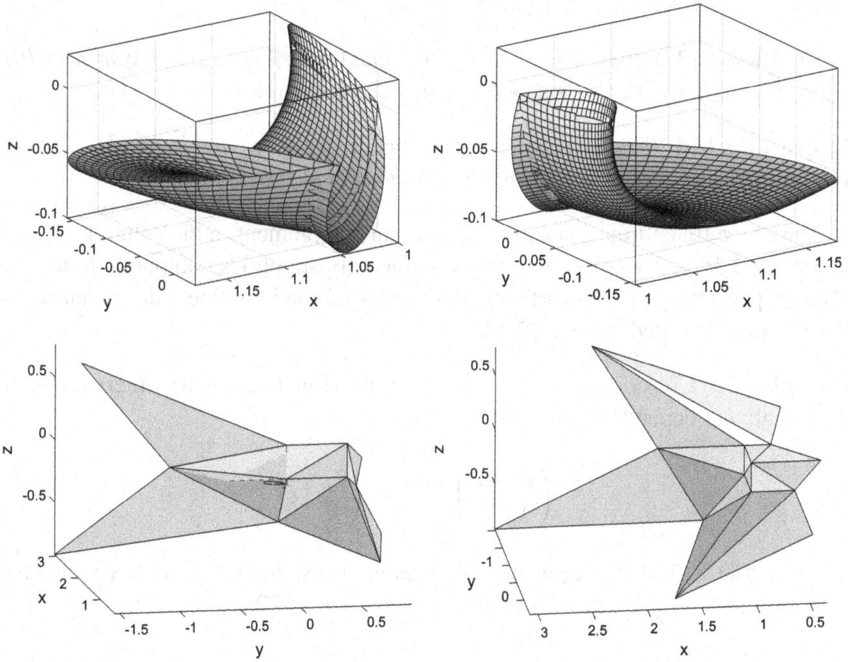

Fig. 4.3 Parametric 3D solution set of system (4.24) shown from different perspectives (top left, top right); number of sample points is 50 for both p_1 and p_2; non-parametric solution set (bottom left, bottom right) with parametric solution set (dark region in bottom left plot)

where $p_1 \in [0, 1]$, $p_2 \in [0, 0.9]$. Figure 4.3 shows the 3D parametric solution set from different perspectives. The surface representing the solution set is obtained by sampling of the parameter intervals. The step size for parameter p_1 is $\delta \approx 0.02$, and for parameter p_2, it is $\delta \approx 0.018$. For the comparison purposes we present in the figure the corresponding non-parametric solution set plotted using INTLAB [232] function `plotlinsol`. As we can see, the difference between the parametric and non-parametric solution set is huge.

4.4 Affine Transformation of System of Linear Equations

Given parametric interval linear system (4.10), the affine transformation (see Sect. 3.3) of $A(\boldsymbol{p})$ and $b(\boldsymbol{p})$ yields a new system with affine linear dependencies:

$$\boldsymbol{C}(e)x = \boldsymbol{c}(e), \tag{4.25}$$

where $\boldsymbol{C}(e)$ is given by (3.39) and $\boldsymbol{c}(e) = \sum_{k=1}^{K} c^{(k)}\varepsilon_k + \boldsymbol{c}$.

Proposition 4.13 *Let system (4.25) be the affine transformation of system (4.10). Then, the solution set of system (4.10) is included in the solution set of system (4.25).*

Proposition 4.14 *If there are only affine-linear dependencies in (4.11), then the solution set of system (4.10) equals the solution set in system (4.25).*

The affine transformation usually causes an enlargement in the solution set (see Examples 4.15 and 4.16); however, it significantly simplifies computation instead. More importantly, it enables PILS with affine-linear and nonlinear dependencies to be treated in a unified manner [259].

Example 4.15 Consider the following two-dimensional parametric interval system with nonlinear dependencies:

$$\begin{pmatrix} p^2 & -2 \\ 1 & p \end{pmatrix} \begin{pmatrix} x \\ y \end{pmatrix} = \begin{pmatrix} 2p \\ 0 \end{pmatrix}, \tag{4.26}$$

where $p \in [1.0, 2.0]$. The affine transformation of system (4.26) yields a parametric interval linear system with affine-linear dependencies (4.25) with

$$\boldsymbol{C}(e) = \begin{pmatrix} 2.375 + 1.5\varepsilon_1 & -2 \\ 1 & 1.5 + 0.5\varepsilon_1 \end{pmatrix} + \begin{pmatrix} [-0.125, 0.125] & 0 \\ 0 & 0 \end{pmatrix}, \tag{4.27a}$$

$$\boldsymbol{c}(e) = \begin{pmatrix} 3 + \varepsilon_1 \\ 0 \end{pmatrix}. \tag{4.27b}$$

The solution sets of system (4.26) and of the corresponding parametric interval system with affine-linear dependencies (4.27) are depicted in Fig. 4.4. The figure also shows the interval hulls of both solution sets.

Example 4.16 Consider the following three-dimensional parametric interval system with nonlinear dependencies:

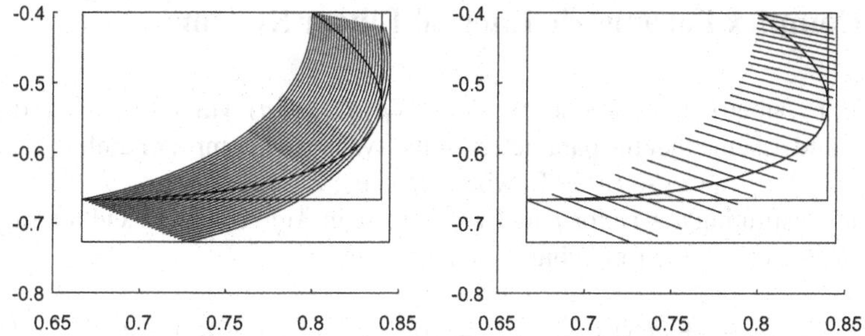

Fig. 4.4 Solution set of system (4.26) (black solid line) and solution set of corresponding system with affine-linear dependencies (set of plane curves depending on parameter p_1, $\delta \approx 0.06$ (left), and set of plane curves depending on parameter p_2, $\delta \approx 0.06$ (right); rectangles represent interval hulls of respective solution sets

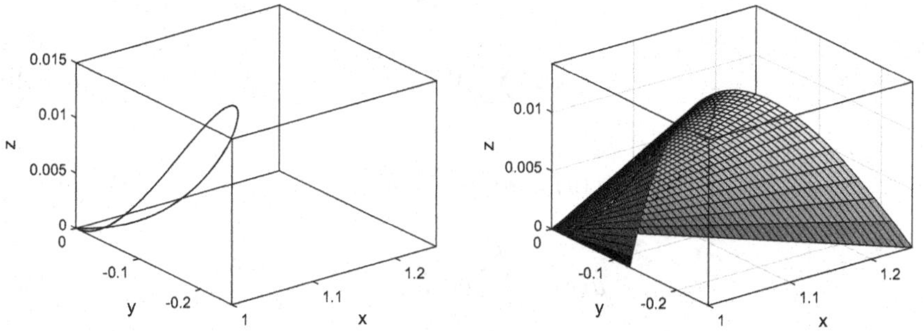

Fig. 4.5 Solution set of system (4.28) (left) and solution set of corresponding system with affine-linear dependencies together with original solution set (right); number of sample points is 50 for both ε_1 and ε_2

$$\begin{pmatrix} 1 & p & p \\ p & 2 & p \\ p & p & 3 \end{pmatrix} \begin{pmatrix} x \\ y \\ z \end{pmatrix} = \begin{pmatrix} 1 \\ p^2 \\ p \end{pmatrix}, \tag{4.28}$$

where $p \in [0, 1]$. The affine transformation of system (4.28) yields parametric interval linear system with affine-linear dependencies (4.25) with

$$\boldsymbol{C}(e) = \begin{pmatrix} 1 & 0.5 + 0.5\varepsilon_1 & 0.5 + 0.5\varepsilon_1 \\ 0.5 + 0.5\varepsilon_1 & 2 & 0.5 + 0.5\varepsilon_1 \\ 0.5 + 0.5\varepsilon_1 & 0.5 + 0.5\varepsilon_1 & 3 \end{pmatrix}, \tag{4.29a}$$

$$\boldsymbol{c}(e) = \begin{pmatrix} 1 \\ 0.375 + 0.5e1 \\ 0.5 + 0.5\varepsilon_1 \end{pmatrix} + \begin{pmatrix} 0 \\ [-0.125, 0.125] \\ 0 \end{pmatrix}. \tag{4.29b}$$

The solution sets of system (4.28) and of transformed system (4.29) are depicted in Fig. 4.5.

4.5 Complex Parametric Interval Linear Systems

Consider parametric interval linear system $A(\boldsymbol{p})x = b(\boldsymbol{p})$ with affine-linear dependencies, and assume that the parameters of the system are complex numbers, i.e., for $k = 1, \dots, K$, $p_k = u_k + iv_k \in \mathbb{C}$, where $u_k \in \boldsymbol{u}_k$, $v_k \in \boldsymbol{v}_k$.

After substituting $u + iv$ for p and $y + iz$ for x in $A(p)x = b(p)$ and rearranging the terms, we obtain the following system:

$$(C(u) + iD(v))(y + iz) = c(u) + id(v),\ u \in \boldsymbol{u}, v \in \boldsymbol{v}, \tag{4.30}$$

where

$$C(u) = A^{(0)} + \sum_{k=1}^{K} A^{(k)} u_k, \tag{4.31a}$$

$$D(v) = \sum_{k=1}^{K} A^{(k)} v_k, \tag{4.31b}$$

$$c(u) = b^{(0)} + \sum_{k=1}^{K} b^{(k)} u_k, \tag{4.31c}$$

$$d(v) = \sum_{k=1}^{K} b^{(k)} v_k. \tag{4.31d}$$

The solution set of system (4.30) is defined in a typical way, i.e.,

$$S(\boldsymbol{u}, \boldsymbol{v}) = \{y + iz \mid \exists u \in \boldsymbol{u}\ \exists v \in \boldsymbol{v}\ :\ (C(u) + iD(v))(y + iz) = c(u) + id(v)\}. \tag{4.32}$$

Obviously, it holds that $S(\boldsymbol{p}) \equiv S(\boldsymbol{u}, \boldsymbol{v})$.

System (4.30) can be transformed to the following real system of parametric interval linear equations (with y, z as unknowns):

$$\begin{cases} C(u)y - D(v)z = c(u) \\ D(v)y + C(u)z = d(v) \end{cases},\ u \in \boldsymbol{u}, v \in \boldsymbol{v}. \tag{4.33}$$

Equivalently, we can write system (4.33) as

$$\begin{pmatrix} C(u) & -D(v) \\ D(v) & C(u) \end{pmatrix} \begin{pmatrix} y \\ z \end{pmatrix} = \begin{pmatrix} c(u) \\ d(v) \end{pmatrix},\ u \in \boldsymbol{u}, v \in \boldsymbol{v}. \tag{4.34}$$

So, the problem of solving a complex parametric interval linear system have been reduced to a problem of solving a real parametric interval linear system. Therefore, all methods described in Chap. 5 can be applied to system (4.30). The main drawback of the proposed approach is that the obtained system is twice as large as the original system.

The open problem that we must address in the future is to develop an RAA-based approach to solving complex parametric interval linear systems with nonlinear dependencies.

4.6 Over- and Under-Determined Parametric Interval Linear Systems

Let $A \in \mathbb{IR}^{m\times n}$ and $b \in \mathbb{R}^m$. If $m > n$, then parametric linear system $Ax = b$ is *over-determined* and, in general, has no solution. Instead, we search for a vector $x \in \mathbb{R}^n$ such that $\|b(p) - A(p)x\|_2$ is minimal. If $m < n$, then the linear system is *under-determined* with, in general, an infinite number of solutions. In such a case, we usually search for a solution with a minimal norm, i.e., we search for a vector $x \in \mathbb{R}^n$ such that $\|x\|_2$ is minimal.

If $\operatorname{rank}(A) = \min\{m, n\}$, then both problems can be solved uniquely; as is well known, this unique solution satisfies one of the following equations:

$$A^T Ax = A^T b,\ m > n \tag{4.35}$$

$$x = A^T y,\ \text{where } AA^T y = b,\ m < n. \tag{4.36}$$

Now, let us consider the parametric interval linear system (4.10) with an $m\times n$-dimensional parametric interval matrix and m-dimensional parametric interval vector. If parametric matrix $A(p)$ has full rank for each $p \in \boldsymbol{p}$, then we can try to solve this system by using the parametric interval version of Eqs. (4.35) and (4.36), i.e.,

$$A(\boldsymbol{p})^T A(\boldsymbol{p})x = A(\boldsymbol{p})^T b(\boldsymbol{p}),\ m > n \tag{4.37}$$

$$x = A(\boldsymbol{p})^T y,\ \text{where } A(\boldsymbol{p})A(\boldsymbol{p})^T y = b(\boldsymbol{p}),\ m < n. \tag{4.38}$$

However, matrices $A(\boldsymbol{p})^T A(\boldsymbol{p})$, $A(\boldsymbol{p})A(\boldsymbol{p})^T$ are, in general, ill-conditioned [226]. Moreover, matrix multiplication makes the dependencies even more complex, in particular, affine-linear dependencies became no longer linear. So, it is better to use instead, as suggested in [226] (cf. [160]), an equivalent form of the least square formula, i.e., to use the following augmented parametric interval linear systems:

$$\begin{pmatrix} A(\boldsymbol{p}) & -I \\ 0 & A(\boldsymbol{p})^T \end{pmatrix} \begin{pmatrix} x \\ y \end{pmatrix} = \begin{pmatrix} b(\boldsymbol{p}) \\ 0 \end{pmatrix},\ m > n \tag{4.39}$$

$$\begin{pmatrix} A(\boldsymbol{p})^T & -I \\ 0 & A(\boldsymbol{p}) \end{pmatrix} \begin{pmatrix} x \\ y \end{pmatrix} = \begin{pmatrix} 0 \\ b(\boldsymbol{p}) \end{pmatrix},\ m < n \tag{4.40}$$

Systems (4.39) and (4.40) are square parametric interval linear systems, hence they can be solved using one of the methods described in Chap. 5. It can be proven easily that if vector $(\boldsymbol{x}, \boldsymbol{y})^T \in \mathbb{IR}^{m+n}$ is an outer interval solution to the system (4.39), then $\boldsymbol{x}$ is an outer interval solution to the corresponding over-determined system. Similarly, if $(\boldsymbol{x}, \boldsymbol{y})^T$ is an outer interval solution to the system (4.40), then $\boldsymbol{y}$ is an outer interval solution to the corresponding under-determined system. As in the case of complex systems, the main drawback of this approach is that we must solve a system of size $(m + n) \times (m + n)$ instead of an $m \times n$-dimensional system. Another drawback (cf. [88]) is that the resulting interval vector contains the solution of the interval least squares problem. This means that the method returns a solution even if the system is unsolvable. An overview of various other approaches to solving over-determined interval linear systems can be found in [88].

Chapter 5
Methods for Solving Parametric Interval Linear Systems

In this chapter we present various methods for solving parametric interval linear systems with general dependencies. Furthermore, we propose several modifications and improvements of selected methods that aim at extending their applicability and accuracy.

However, before we proceed to the description of the methods, we must explain what we really mean by claiming "to solve" a parametric interval linear system. As we saw in Sect. 4.3, a parametric (united) solution set has, in the general case, a very complex structure. Therefore, in practical applications, we are usually interested in an *interval solution* to a parametric interval linear system, i.e., we are interested in an interval vector that encloses the (united) parametric solution set. Generally, we can distinguish the following interval solutions:

- *outer interval (OI) solution*

$$\boldsymbol{x}^{OI} \supseteq S(\boldsymbol{p}), \tag{5.1}$$

- *interval hull (IH) solution*

$$\boldsymbol{x}^{IH} \equiv \text{hull}\,(S(\boldsymbol{p}))\,, \tag{5.2}$$

- *inner estimate of the hull (IEH) solution*

$$\boldsymbol{x}^{IEH} \subseteq \text{hull}\,(S(\boldsymbol{p}))\,. \tag{5.3}$$

Computing the interval hull for a solution set of interval linear systems has been proven to be NP-hard (see, e.g., [126, 214, 220]). So, the analogue problem for parametric interval linear systems (which is even more general) is NP-hard as well. Thus, every general purpose algorithm (which assumes only regularity of the system matrix) requires, in the worst case, a number of operations that is exponential in the number of equations. Polynomial time algorithms are possible for systems that fulfill

I. Skalna, *Parametric Interval Algebraic Systems*, Studies in Computational Intelligence 766, https://doi.org/10.1007/978-3-319-75187-0_5

some special conditions (such as the monotonicity of the solution with respect to all parameters) or have a specific structure.

For these reasons, most of the existing methods for solving PILS are concerned with determining approximate solutions. Obviously, the goal is to obtain an outer interval solution (hopefully as tight as possible) or an interval estimate to the hull solution (hopefully as wide as possible). Jansson [90] was among the first to consider interval systems with a special structure. He developed a method based on fixed-point iteration for solving interval linear systems with symmetric and skew-symmetric matrices as well as dependencies in the right-hand side. Rump [229] proposed a generalization of Jansson's approach to arbitrary affine-linear dependencies of the matrix and the right-hand-side vector on a set of parameters (cf. [172–174, 245, 246, 248]). Rump's method was then investigated by Skalna (see, e.g., [247]) and Popova (see, e.g., [173]) and implemented by Popova and Krämer [189]. Other iterative methods for computing OI solutions were proposed by Kolev [107, 116], and El-Owny [48]. The so-called direct methods were given by Skalna [248] and Kolev [109, 113]. A monotonicity approach was investigated by Kolev [105], Popova [176], Rohn [215], and Skalna [250]. Parametric versions of the Bauer–Skeel and Hansen–Bliek–Rohn methods were discussed by Hladík [80]. Degrauwe et al. [36] developed a method based on a Neumann series. Affine-Interval Gaussian Elimination was investigated by Akhmerov [2]. The problem of computing the interval hull solution was considered by Kolev [108, 112, 113], and Skalna [252], for example. There are also a few methods for computing an inner estimate of the hull (IEH) solution (see, e.g., [113, 192, 229, 249, 256]).

A more-general approach to the problem of solving PILS was developed by Kolev [113, 116], who introduced a new type of solution called a *parametric solution* or a *p-solution*, which is of the following parametric form:

$$\boldsymbol{x}(p) = Lp + \boldsymbol{a}, \tag{5.4}$$

where $L \in \mathbb{R}^{n\times K}$ and $\boldsymbol{a} \in \mathbb{IR}^n$. The p-solution has many useful properties [113]. In particular, it permits us to compute both the OI and IEH solutions as well as what follows the boxes containing the lower and upper bounds of the IH solution. However, the main advantage of the parametric solution is that it can be laid as a basis for a new paradigm for solving various problems related to parametric interval linear systems, such as the following optimization problem [113]: find global minimum

$$g^* = \min g(x, p) \tag{5.5}$$

subject to constraints $A(p)x(p) = b(p)$, $p \in \boldsymbol{p}$, where $g(x, p)$ is, in general, a non-linear function [113].

5.1 Methods for Approximate Solutions

In this section, we discuss methods for computing approximate solutions to parametric interval linear systems, i.e., methods for computing OI and IEH solutions.

5.1.1 *Modified Rump's Fixed-Point Iteration*

In [226], Rump proposed a verification method for computing an enclosure for the solution of systems of equations known as *Rump's Fixed-point Iteration* (RFPI) (or Krawczyk iteration). The main idea of the method is to verify that a certain interval is mapped by the Krawczyk operator $K(x) = \tilde{x} + R(b - A\tilde{x}) + (I - RA)x$ into itself or into its interior. The theoretical justification for the method is given by the following theorem:

Theorem 5.1 (Rump [226]) *Let* $\mathbf{A} \in \mathbb{IR}^{n\times n}$, $\mathbf{b} \in \mathbb{IR}^n$, $R \in \mathbb{R}^{n\times n}$, *and* $\tilde{x} \in \mathbb{R}^n$. *If, for some* $\mathbf{x} \in \mathbb{IR}^n$

$$R(\mathbf{b} - \mathbf{A}\tilde{x}) + (I - R\mathbf{A})\mathbf{x} \subseteq \operatorname{int}(\mathbf{x}), \tag{5.6}$$

then the following is true for each $A \in \mathbf{A}$ *and each* $b \in \mathbf{b}$: *A and R are non-singular, and there is one and only one* $x \in \tilde{x} + \operatorname{int}(\mathbf{x})$ *such that* $Ax = b$.

Jansson [90] introduced a modification to Rump's approach that enabled the dependencies in the data to be taken into account. He considered *symmetric* and *skew-symmetric* matrices as well as dependencies in the right-hand vector. The result of Jansson was then generalized by Rump [229] to parametric interval linear systems, with a particular emphasis on parametric interval linear systems with affine-linear dependencies.

Theorem 5.2 (cf. Rump [229]) *Let* $A(\mathbf{p})x = b(\mathbf{p})$ *be a parametric interval linear system with affine-linear dependencies,* $R \in \mathbb{R}^{n\times n}$, $\mathbf{y} \in \mathbb{IR}^n$, $\tilde{x} \in \mathbb{R}^n$, *and define* $\mathbf{z} \in \mathbb{IR}^n$ *and* $\mathbf{G} \in \mathbb{IR}^{n\times n}$ *by*

$$\mathbf{z} = R\left(b^{(0)} - A^{(0)}\tilde{x}\right) + \sum_{k=1}^{K}\left(R\left(b^{(k)} - A^{(k)}\tilde{x}\right)\right)\mathbf{p}_k, \tag{5.7a}$$

$$\mathbf{G} = I - R\mathbf{A}, \tag{5.7b}$$

where $\mathbf{A}$ *is the value of an interval extension of* $A(p)$ *over* $\mathbf{p}$. *Define* $\mathbf{v} \in \mathbb{IR}^n$ *by means of the following single-step formula*[1]:

$$\mathbf{v}_i = \{\mathbf{z} + \mathbf{G}\mathbf{u}\}_i, \quad i = 1, \ldots, n, \tag{5.8}$$

[1] In the original version of the theorem, Rump used the German term *Einzelschrittverfahren*, which means "single-step"; so, the general idea of the theorem is preserved.

where $\boldsymbol{u} = (\boldsymbol{v}_1, \ldots, \boldsymbol{v}_{i-1}, \boldsymbol{y}_i, \ldots, \boldsymbol{y}_n)^T$.

If $\boldsymbol{v} \subseteq \mathrm{int}(\boldsymbol{y})$, *then* R *and each matrix* $A(p)$, $p \in \boldsymbol{p}$, *is non-singular, and for each* $p \in \boldsymbol{p}$, *unique solution* $x = A(p)^{-1}b(p)$ *of system* $A(p)x = b(p)$ *satisfies* $x \in \tilde{x} + \boldsymbol{v}$, *i.e.,* $S(\boldsymbol{p}) \subseteq \tilde{x} + \boldsymbol{v}$. *Additionally, with* $\boldsymbol{d} = \boldsymbol{G}\boldsymbol{v}$, *we have, for* $i = 1 \ldots, n$,

$$\tilde{x}_i + [\underline{z}_i + \overline{d}_i, \overline{z}_i + \underline{d}_i] \subseteq \mathrm{hull}\,(S(\boldsymbol{p}))_i\ .$$

In (Skalna [247]) (cf. [174]), we have proposed an important modification of RFPI that has significantly extended the applicability of this method (see Example 5.4). Our modification consisted of replacing matrix $\boldsymbol{G}$ in (5.7b) with matrix

$$\boldsymbol{G}' = \mathrm{hull}\,(\{I - RA(p) \mid p \in \boldsymbol{p}\}) = I - RA^{(0)} + \sum_{k=1}^{K} \left(RA^{(k)}\right) \boldsymbol{p}_k, \tag{5.9}$$

i.e., with the matrix computed by taking into account the dependencies in the data. The last equality in (5.9) follows from Theorem 3.33.

For the purposes of this book, the improved version of RFPI will be referred to as the *Modified Rump's Fixed-Point Iteration* (MRFPI).

Theorem 5.3 *Let* $A(\boldsymbol{p})x = b(\boldsymbol{p})$ *be a parametric interval linear system with affine-linear dependencies,* $R \in \mathbb{R}^{n\times n}$, $\boldsymbol{y} \in \mathbb{IR}^n$, $\tilde{x} \in \mathbb{R}^n$. *Let* $\boldsymbol{z} \in \mathbb{IR}^n$ *be given by formula (5.7a) and* $\boldsymbol{G}' \in \mathbb{IR}^{n\times n}$ *by formula (5.9). Define* $\boldsymbol{w} \in \mathbb{IR}^n$ *by*

$$\boldsymbol{w}_i = \{\boldsymbol{z} + \boldsymbol{G}'\boldsymbol{u}\}_i, \quad i = 1, \ldots, n, \tag{5.10}$$

$\boldsymbol{u} = (\boldsymbol{w}_1, \ldots, \boldsymbol{w}_{i-1}, \boldsymbol{y}_i, \ldots, \boldsymbol{y}_n)^T$. *If* $\boldsymbol{w} \subseteq \mathrm{int}(\boldsymbol{y})$, *then* R *and each matrix* $A(p)$, $p \in \boldsymbol{p}$, *is non-singular, and for each* $p \in \boldsymbol{p}$, *unique solution* $x = A(p)^{-1}b(p)$ *of system* $A(p)x = b(p)$ *satisfies* $x \in \tilde{x} + \boldsymbol{w}$, *i.e.,* $S(\boldsymbol{p}) \subseteq \tilde{x} + \boldsymbol{w}$. *Additionally, with* $\boldsymbol{d} = \boldsymbol{G}'\boldsymbol{w}$, *we have*

$$\tilde{x}_i + [\underline{z}_i + \overline{d}_i, \overline{z}_i + \underline{d}_i] \subseteq \mathrm{hull}\,(S(\boldsymbol{p}))\ ,\ i = 1 \ldots, n.$$

Proof The theorem follows from Theorem 2.1 in [229] and Theorem 3.33. □

Obviously, $\boldsymbol{G}' \subseteq \boldsymbol{G}$, and hence $\boldsymbol{w} \subseteq \boldsymbol{v}$, which means that the MRFPI is never worse than RFPI. Moreover, by Proposition 3.5, $\rho\left(\left|\boldsymbol{G}'\right|\right) \leqslant \rho(|\boldsymbol{G}|)$. Since RFPI and the MRFPI converge if, respectively, $\rho(|\boldsymbol{G}|) < 1$ and $\rho\left(\left|\boldsymbol{G}'\right|\right) < 1$, the MRFPI should perform much better. Using a simple example, we show how useful our improvement is.

Example 5.4 (cf. [174]) Consider an n-dimensional parametric interval matrix with

$$A(p) = \begin{cases} p_j, & i \leqslant j, \\ 0, & i = j + 2, \\ 1, & \text{otherwise} \end{cases}$$

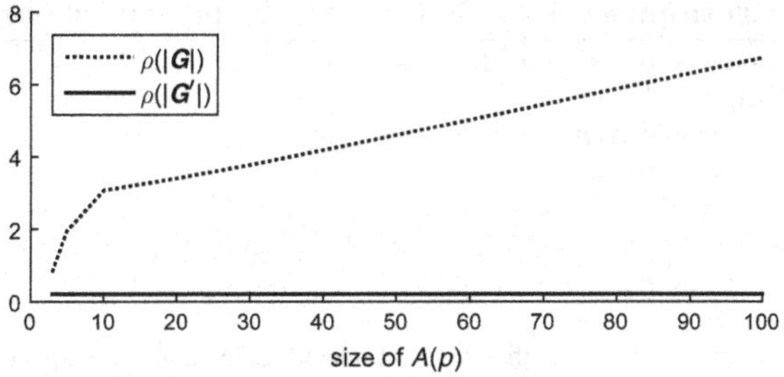

Fig. 5.1 Comparison of $\rho(|\boldsymbol{G}|)$ and $\rho(|\boldsymbol{G}'|)$ for Example 5.4

and $p_j \in (j+1)\cdot[0.9, 1.1]$. Spectral radius $\rho(|\boldsymbol{G}'|) = 0.2$ regardless of the size of the matrix, whereas spectral radius $\rho(|\boldsymbol{G}|)$ increases along with n (see Fig. 5.1). As we can see, for systems of a size larger than 5, the convergence condition of RFPI is no longer fulfilled.

The applicability of the MRFPI can be extended to parametric interval linear systems with non-linear dependencies in two ways. We can transform the system using the affine transformation from Sect. 4.4; then, Theorem 5.3 applies directly. Or, we can combine the following theorem with a method for range bounding (cf. Sect. 1.3):

Theorem 5.5 *Let $A(\boldsymbol{p})x = b(\boldsymbol{p})$ be a parametric interval linear system, $R \in \mathbb{R}^{n\times n}$, $\boldsymbol{y} \in \mathbb{IR}^n$, $\tilde{x} \in \mathbb{R}^n$. Let $z(p) \in \mathbb{R}^n$ and $G(p) \in \mathbb{R}^{n\times n}$ be given by*

$$z(p) = R\left(b(p) - A(p)\tilde{x}\right), \tag{5.11a}$$
$$G(p) = I - RA(p), \tag{5.11b}$$

and let $\boldsymbol{z}$ and $\boldsymbol{G}$ be the values on interval extensions of $z(p)$ and $G(p)$ over $\boldsymbol{p}$, respectively. Define $\boldsymbol{w} \in \mathbb{IR}^n$ by

$$\boldsymbol{w}_i = \{\boldsymbol{z} + \boldsymbol{G}\boldsymbol{u}\}_i, \quad i = 1, \ldots, n, \tag{5.12}$$

$\boldsymbol{u} = (\boldsymbol{w}_1, \ldots, \boldsymbol{w}_{i-1}, \boldsymbol{y}_i, \ldots, \boldsymbol{y}_n)^T$. If $\boldsymbol{w} \subseteq \operatorname{int}(\boldsymbol{y})$, then R and each matrix $A(p)$, $p \in \boldsymbol{p}$, is non-singular, and for each $p \in \boldsymbol{p}$, unique solution $x = A(p)^{-1}b(p)$ of system $A(p)x = b(p)$ satisfies $x \in \tilde{x} + \boldsymbol{w}$, i.e., $S(\boldsymbol{p}) \subseteq \tilde{x} + \boldsymbol{w}$. Additionally, with $\boldsymbol{d} = \boldsymbol{G}\boldsymbol{w}$, we have

$$\tilde{x}_i + [\underline{z}_i + \overline{d}_i, \overline{z}_i + \underline{d}_i] \subseteq \operatorname{hull}\left(S(\boldsymbol{p})\right), \; i = 1\ldots, n.$$

Remark The proof of Theorem 5.5 follows along the lines of the analogous theorems from [229], therefore, it is omitted.

In order to obtain sharp bounds on the parametric solution set, the ranges on interval extensions $\boldsymbol{z}$ and $\boldsymbol{G}$ should be as tight as possible.

Algorithm 4 presents a procedure developed based on the result formulated in Theorem 5.5. Operation $\boldsymbol{w} \circ \epsilon$, called *epsilon-inflation* [225], aims to increase the

Algorithm 4 (Modified Rump's Fixed-point Iteration)

Input: $A(\boldsymbol{p})x = b(\boldsymbol{p})$, $\boldsymbol{p}$, ϵ, $k_{\max}$
Output: $\boldsymbol{x}^{OI}$ and $\boldsymbol{x}^{IEH}$

```
  A^c = mid(A(p))
  b^c = mid(b(p))
  R ≈ (A^c)^{-1}
  x^c = R b^c
  z(p) := R(b(p) − A(p)x^c)
  G(p) := I − RA(p)
  Compute z and G (the values of interval extensions of, respectively, z(p) and G(p) over p) using
  a preferred method for range bounding
  w = z;  k = 0 // k denotes the number of iterations
  repeat
    k = k + 1
    y = w = w ∘ ε // epsilon-inflation
    for i = 1 : n do
      w_i = {z + Gw}_i
    end for
  until w ⊆ int(y) or k > k_max
  if w ⊆ int(y) then
    x^{OI} = x^c + w;
    d = Gw;  x^{IEH} = x^c + [inf(z(p)) + d̄, sup(z(p)) + d̲]
    return "Success"
  else
    return "Failure"
  end if
```

chances for inclusion $\boldsymbol{w} \subseteq \operatorname{int}(\boldsymbol{y})$ to be satisfied (cf. [24, 228, 231]). Epsilon-inflation can be defined in various ways [231]. In our computer implementation, we use the following formula:

$$\boldsymbol{x} \circ \epsilon = \boldsymbol{x} + x^{\Delta}[-\epsilon, \epsilon] + [-\eta, \eta], \tag{5.13}$$

where $\epsilon \geqslant 0$ and η denoted denotes the smallest representable positive machine number. The epsilon-inflation defined by (5.13) guarantees the validity of the following assertions [231]:

(i) If $\rho(|\boldsymbol{G}|) < 1/(1 + 2\epsilon)$, then inclusion $\boldsymbol{w} \subseteq \boldsymbol{y}$ is satisfied for some finite k.
(ii) If $\boldsymbol{w} \subseteq \boldsymbol{y}$ is satisfied for some finite k, then $\rho(|\boldsymbol{G}|) < 1$.

If $\epsilon = 0$, then (i) and (ii) are equivalent.

The most-time-consuming part of the algorithm is the computation of $\boldsymbol{z}$ and $\boldsymbol{G}$. If there are only affine-linear dependencies in $A(p)$ and $b(p)$, then the asymptotic time complexity of the MRFPI is $\mathcal{O}(Kn^3 + \kappa n^2)$, where κ is the number of iterations. In the general case, the computational complexity will depend on the method used to bound the range of $z(p)$ and $(G(p)$ over $\boldsymbol{p}$. Obviously, the computational time also depends on the number of iterations, whose number, in turn, depends on the choice of η and $\epsilon \geqslant 0$. According to Rump [231], 0.1 and 0.2 are reasonable values for ϵ in (5.13) (generally, ϵ should not be greater than 0.5). With these values, the number of iterations is usually less than 10. A broader discussion on the epsilon-inflation and convergence behavior of RFPI can be found in [227, 231].

In our implementation of the MRFPI, we use revised affine forms to perform computations on parametric vectors and matrices. Regarding other approaches, Popova (see, e.g., [178]) combined RFPI with the arithmetic of generalized (proper and improper) intervals to solve systems with rational parameter dependency. However, this methodology is not efficient for non-monotone rational functions [60]. Garloff et al. [60] combined RFPI with Bernstein's expansion of multivariate polynomials to solve systems with polynomial and rational parameter dependency. El-Owny [45] used Hansen's generalized arithmetic to bound the ranges of $z(p)$ and $G(p)$ over $\boldsymbol{p}$, which enabled a more-complex non-linear dependencies to be taken into account.

5.1.2 Parametric Gauss–Seidel Iteration

In this section, we describe an iterative method, which is usually used to iteratively improve initial interval enclosure $\boldsymbol{x}$ of parametric solution set $S(\boldsymbol{p})$. In order to achieve better convergence properties, system $A(\boldsymbol{p})x = b(\boldsymbol{p})$ is first preconditioned with some non-singular matrix R, usually the numerically computed inverse of mid-point matrix $A(p^c)$. So, let us consider following pre-conditioned system

$$C(\boldsymbol{p})x = c(\boldsymbol{p}),$$

and let $\boldsymbol{y}_j$ $(j = 1, \dots, n)$ be an interval enclosure for

$$\left\{ \frac{c(p)_j - \sum_{k=1}^{j-1} L(p)_{jk} y_k - \sum_{k=j+1}^{n} U(p)_{jk} x_k}{D_{jj}(p)}, x \in \boldsymbol{x}, y \in \boldsymbol{y}, p \in \boldsymbol{p} \right\}, \tag{5.14}$$

where the components of $D(p)$, $L(p)$, and $U(p)$ are defined by

$$L_{ij}(p) = \begin{cases} C(p)_{ij}, & i > j, \\ 0, & i \leqslant j, \end{cases} \quad U_{ij}(p) = \begin{cases} C(p)_{ij}, & i < j, \\ 0, & i \geqslant j, \end{cases}$$

$$D_{ij}(p) = \begin{cases} C(p)_{ii}, & i = j, \\ 0, & i \neq j. \end{cases}$$

If $\boldsymbol{y}$ is strictly included in $\boldsymbol{x}$, we can hope to obtain further improvement of the enclosure of $S(\boldsymbol{p})$ by repeating the above described procedure. This leads to the following iterative scheme (cf. [160]):

$$\boldsymbol{x}^0 = \boldsymbol{x}, \tag{5.15}$$

$$\boldsymbol{x}^{i+1} = \boldsymbol{y}^i \cap \boldsymbol{x}^i, \ (i = 1, 2, \dots), \tag{5.16}$$

which we will refer to as *Parametric Interval Gauss–Seidel Iteration* (PIGSI) method (cf. [83]). The sketch of the algorithm of the PIGSI method is presented in Algorithm 5. The asymptotic time complexity of the algorithm is $\mathcal{O}(n^3 + n^2\kappa\tau)$, where κ is the cost of computing the value of $\boldsymbol{y}^i_j$; and τ is the number of iterations.

Algorithm 5 (Parametric Interval Gauss–Seidel Iteration)

Input: $A(\boldsymbol{p})x = b(\boldsymbol{p})$, $\boldsymbol{p}$, $\boldsymbol{x}$ (verified initial enclosure of $S(\boldsymbol{p})$), ϵ
Output: $\boldsymbol{x}^{OI}$
  $A^c = \text{mid}(A(\boldsymbol{p}))$
  $R \approx (A^c)^{-1}$
  $C(p) = RA(p)$
  $c(p) = Rb(p)$
  if $\left(\rho\left(\left|(D(\boldsymbol{p}) + L(\boldsymbol{p}))^{-1}\, U(\boldsymbol{p})\right|\right) < 1\right)$ **then**
    $\boldsymbol{x}^0 = \boldsymbol{x}$ // starting point
    $i = -1$
    repeat
      $i = i + 1$
      for $j = 1$ to n **do**
        Compute $\boldsymbol{y}^i_j$, interval enclosure for
        $\left\{\frac{c_j(p) - \sum_{k=1}^{j-1} L_{jk}(p)y^i_k - \sum_{k=j+1}^{n} U_{jk}(p)x^i_k}{D_{jj}(p)},\, x \in \boldsymbol{x}^i,\, y \in \boldsymbol{y}^i,\, p \in \boldsymbol{p}\right\}$
      end for
      $\boldsymbol{x}^{i+1} = \boldsymbol{y}^i \cap \boldsymbol{x}^i$
    until $\left(q\left(\boldsymbol{x}^{i+1}, \boldsymbol{x}^i\right) < \epsilon\right)$
    $\boldsymbol{x}^{OI} = \boldsymbol{x}^{i+1}$
    return "Success"
  else
    return "Failure"
  end if

The distance between interval vectors $\boldsymbol{x}$ and $\boldsymbol{y}$, which is used in the stopping criterion of the PIGSI, is assessed here using the following formula:

$$q(\boldsymbol{x}, \boldsymbol{y}) = \max\left\{\max_{j=1,\ldots,n}\left|\underline{x}_j - \underline{y}_j\right|,\ \max_{j=1,\ldots,n}\left|\overline{x}_j - \overline{y}_j\right|\right\}. \tag{5.17}$$

Parametric Interval Gauss–Seidel iteration can also be useful when we want to improve the enclosure for the set of those solution that lie in a specific initial box, i.e., if we want to improve enclosure for the *truncated solution set* $S(\boldsymbol{p}) \cap \boldsymbol{x}^0$. It is worth to note that the truncated solution set is always bounded, even if solution set $S(\boldsymbol{p})$ itself is not bounded. This means that the PIGSI can be applied to non-singular system, however, as mentioned in [160], with limited success only.

5.1.3 Direct Method

In (Skalna [248]), we proposed the so-called *Direct Method* (DM) for solving parametric interval linear systems with affine-linear dependencies. Below, we present a theorem that generalizes Theorem 3 from [248] to parametric interval linear systems with arbitrary dependencies.

Theorem 5.6 *Let $A(\boldsymbol{p})x = b(\boldsymbol{p})$ be a parametric interval linear system, R some real matrix, and $\tilde{x}$ some real vector. Let $z(p)$ be given by (5.11a) and define $B(p) \in \mathbb{R}^{n\times n}$ by*

$$B(p) = RA(p). \tag{5.18}$$

Let $\boldsymbol{z}$ and $\boldsymbol{B}$ be the interval extensions of $z(p)$ and $B(p)$ over $\boldsymbol{p}$, respectively. If $\boldsymbol{B}$ is an H-matrix, then

$$S(\boldsymbol{p}) \subseteq \tilde{x} + \langle \boldsymbol{B}\rangle^{-1}|\boldsymbol{z}|[-1,\ 1]. \tag{5.19}$$

Proof If $x \in S(\boldsymbol{p})$, then there is $p \in \boldsymbol{p}$ such that $A(p)x = b(p)$. Since $\boldsymbol{B}$ is an H-matrix, hence $A(p)$ is non-singular, and we have

$$\begin{aligned} x = A(p)^{-1}b(p) &= \tilde{x} + A(p)^{-1}\left(b(p) - A(p)\tilde{x}\right) = \\ &= \tilde{x} + (RA(p))^{-1}\left(R(b(p) - A(p)\tilde{x})\right). \end{aligned}$$

Since $RA(p) \in \boldsymbol{B}$ and $R(b(p) - A(p)\tilde{x}) \in \boldsymbol{z}$, hence $u = x - \tilde{x}$ is a solution to interval linear system $\boldsymbol{B}u = \boldsymbol{z}$. This means that (Theorem 3.7.7, [160])

$$x \in \tilde{x} + S(\boldsymbol{B}, \boldsymbol{z}) \subseteq \tilde{x} + \langle \boldsymbol{B}\rangle^{-1}|\boldsymbol{z}|[-1,\ 1],$$

which completes the proof. □

The modification of the right-hand side, which relies on substituting $b(p)$ with $b(p) - A(p)\tilde{x}$ and which will be referred to as *residual correction*, aims to narrow the enclosures produced by the Direct Method. As pointed out in [160], given an interval linear system $\boldsymbol{A}x = \boldsymbol{b}$, if right-hand side $\boldsymbol{b}$ is of order $\mathcal{O}(\epsilon)$, then enclosure $\langle \boldsymbol{A}\rangle^{-1}|\boldsymbol{b}|[-1, 1]$ is also of order $\mathcal{O}(\epsilon)$. So, we find solution $\tilde{x}$ to particular system $A(p)x = b(p)$, $p \in \boldsymbol{p}$, and then add to $\tilde{x}$ an enclosure for the solution set of the system with the right-hand vector modified by using the residual correction.

The next proposition shows that for systems with affine-linear dependencies the best choice for R is the numerically computed inverse of the mid-point matrix. The theorem generalizes Theorem 4.1.10 from [160].

Proposition 5.7 *Let $A(\boldsymbol{p}) \subseteq \mathbb{R}^{n\times n}$ be strongly regular. Then, for all $b(\boldsymbol{p}) \subseteq \mathbb{R}^n$ and all $R \in \mathbb{R}^{n\times n}$ such that $\boldsymbol{C} = \mathrm{hull}\,(\{RA(p) \mid p \in \boldsymbol{p}\})$ is an H-matrix, we have*

$$\langle \boldsymbol{B}\rangle^{-1}|\boldsymbol{z}| \leqslant \langle \boldsymbol{C}\rangle^{-1}|\boldsymbol{w}|, \tag{5.20}$$

where $\boldsymbol{B} = \text{hull}\left(\{A(p^c)^{-1}A(p) \mid p \in \boldsymbol{p}\}\right)$, $\boldsymbol{z} = \text{hull}\left(\{A(p^c)^{-1}b(p) \mid p \in \boldsymbol{p}\}\right)$, *and* $\boldsymbol{w} = \text{hull}\left(\{Rb(p) \mid p \in \boldsymbol{p}\}\right)$.

Proof Since $A(\boldsymbol{p})$ is strongly regular, hence $\boldsymbol{B}$ is an H-matrix, and by Theorem 3.13, $\langle \boldsymbol{B}\rangle^{-1} \geqslant 0$. Using the same reasoning as in the proof of Theorem 3.33, it can be shown that

$$\boldsymbol{z} = A(p^c)^{-1}b^{(0)} + \sum_{k=1}^{K} \left|A(p^c)^{-1}b^{(k)}\right| p_k^{\Delta}[-1, 1],$$

$$\boldsymbol{w} = Rb^{(0)} + \sum_{k=1}^{K} \left|Rb^{(k)}\right| p_k^{\Delta}[-1, 1].$$

Then, we have

$$\begin{aligned} |\boldsymbol{z}| &= \left|A(p^c)^{-1}b^{(0)} + \sum_{k=1}^{K} \left|A(p^c)^{-1}b^{(k)}\right| p_k^{\Delta}[-1, 1]\right| = \\ &= \left|(RA(p^c))^{-1}Rb^{(0)} + \sum_{k=1}^{K} \left|(RA(p^c))^{-1}Rb^{(k)}\right| p_k^{\Delta}[-1, 1]\right| \leqslant \\ &\leqslant \left|(RA(p^c))^{-1}\right| |\boldsymbol{w}| \leqslant \langle RA(p^c)\rangle^{-1}|\boldsymbol{w}|. \end{aligned}$$

The last inequality holds true, since $\boldsymbol{C}$ is an H-matrix (Theorem 3.17(iii)). Moreover,

$$\begin{aligned} \langle \boldsymbol{B}\rangle &= I - B^{\Delta} = I - \sum_{k=1}^{K} \left|A(p^c)^{-1}A^{(k)}\right| p_k^{\Delta} = \\ &= I - \sum_{k=1}^{K} \left|(RA(p^c))^{-1}RA^{(k)}\right| p_k^{\Delta} \geqslant I - \langle RA(p^c)\rangle^{-1}C^{\Delta} = \\ &= \langle RA(p^c)\rangle^{-1}(\langle RA(p^c)\rangle - C^{\Delta}) = \langle RA(p^c)\rangle^{-1}\langle \boldsymbol{C}\rangle \end{aligned}$$

So,

$$\begin{aligned} |\boldsymbol{z}| &\leqslant \langle RA(p^c)\rangle^{-1}|\boldsymbol{w}| = \langle RA(p^c)\rangle^{-1}\langle \boldsymbol{C}\rangle\langle \boldsymbol{C}\rangle^{-1}|\boldsymbol{w}| \leqslant \\ &\leqslant \langle \boldsymbol{B}\rangle\langle \boldsymbol{C}\rangle^{-1}|\boldsymbol{w}|. \end{aligned}$$

Multiplication of the above by $\langle \boldsymbol{B}\rangle^{-1} \geqslant 0$ gives (5.20). □

The asymptotic time complexity of the Direct Method is $\mathcal{O}(Kn^3)$. As shown in Algorithm 6, the Direct Method involves operations on real vectors and matrices. Since the method does not have a built-in verification mechanism, one can use of a more expensive version of the algorithm where all operations are performed on interval quantities.

Algorithm 6 (Direct Method)

Input: $A(\boldsymbol{p})x = b(\boldsymbol{p}), \boldsymbol{p}$
Output: $\boldsymbol{x}^{OI}$
  $A^c = \text{mid}(A(\boldsymbol{p}))$
  $b^c = \text{mid}(b(\boldsymbol{p}))$
  $R \approx (A(p^c))^{-1}$
  $x^c = Rb^c$
  $B(p) := RA(p)$
  $z(p) := R(b(p) - A(p)x^c)$ // residual correction
  Compute $\boldsymbol{z}$ and $\boldsymbol{B}$ using a preferred method for range bounding
  if $(\rho(B^{\Delta}) < 1)$ **then**
    $\boldsymbol{x}^{OI} = x^c + \langle \boldsymbol{B} \rangle^{-1}|\boldsymbol{z}|[-1, 1]$
    return "Success"
  else
    return "Failure"
  end if

Recently, parametric interval linear systems with a multiple right-hand side have been considered in [37, 185]. They are of the following form:

$$A(\boldsymbol{p})X = B(\boldsymbol{p}), \tag{5.21}$$

where $X \in \mathbb{R}^{n \times m}$, and $B(\boldsymbol{p}) = (b_1(\boldsymbol{p}) \ldots b_m(\boldsymbol{p}))$ is an $n \times m$ parametric interval matrix. The Direct Method can be straightforwardly extended to solve such systems.

Theorem 5.8 *Let $A(\boldsymbol{p})x = B(\boldsymbol{p})$ be an n-dimensional parametric interval linear system, $R \in \mathbb{R}^{n \times n}$, $\tilde{X} \in \mathbb{R}^{n \times m}$, and $\boldsymbol{p} \in \mathbb{IR}^K$. Define $Z(p) \in \mathbb{R}^{n \times m}$ and $C(p) \in \mathbb{R}^{n \times n}$ by*

$$Z(p) = R\left(B(p) - A(p)\tilde{X}\right), \tag{5.22a}$$

$$C(p) = RA(p), \tag{5.22b}$$

and let $\boldsymbol{Z}$ and $\boldsymbol{C}$ be the interval extensions of $Z(p)$ and $C(p)$ over $\boldsymbol{p}$, respectively. If $\boldsymbol{C}$ is an H-matrix, then

$$S(\boldsymbol{p}) \subseteq \tilde{X} + \langle \boldsymbol{C} \rangle^{-1}|\boldsymbol{Z}|[-1,\ 1]. \tag{5.23}$$

The proof of the above theorem is similar to the proof of Theorem 5.6, therefore it is omitted. The extended version of the Direct Method, for systems with affine-linear dependencies, has the asymptotic time complexity $\mathcal{O}(n^3K + n^2mK)$. The time complexity in the general case will depend on a method used to compute $\boldsymbol{C}$ and $\boldsymbol{z}$.

Example 5.9 Consider parametric interval linear system

$$\begin{pmatrix} 2p_1 & p_2 \\ -p_2 & 2p_1 \end{pmatrix} \begin{pmatrix} x_1 \\ x_2 \end{pmatrix} = \begin{pmatrix} 0 & p_3 \\ p_3 & 0 \end{pmatrix}, \tag{5.24}$$

where $p_1, p_2, p_3 \in [0.9, 1.1]$. The extended version of the Direct Method yields the following outer interval solution

$$\boldsymbol{x}^{OI} = \begin{pmatrix} [-0.27964, -0.120358] & [0.304031, 0.495968] \\ [0.304031, 0.495968] & [0.120358, 0.279642] \end{pmatrix}.$$

For the comparison purposes, we present the result of the modified Krawczyk method from [37] (obtained after 2 iterations):

$$\boldsymbol{x}^{OI} = \begin{pmatrix} [-0.2807, -0.1193] & [0.3029, 0.4971] \\ [0.3029, 0.4971] & [0.1193, 0.2807] \end{pmatrix}.$$

5.1.4 Generalized Parametric Bauer–Skeel Method

The non-parametric version of the *Bauer–Skeel* (BS) method is based on the following theorem (cf. [217]).

Theorem 5.10 (Bauer [14], Skeel [260]) *Let $\boldsymbol{A}x = \boldsymbol{b}$ be an interval linear system. If spectral radius $\rho\left(\left|(A^c)^{-1}\right| A^{\Delta}\right) < 1$, then each $A \in \boldsymbol{A}$ is non-singular and*

$$S(\boldsymbol{A}, \boldsymbol{b}) \subseteq x^c + (x^* - |x^c|)[-1, 1],$$

where

$$\begin{aligned} x^c &= (A^c)^{-1} b^c, \\ M &= \left(I - \left|(A^c)^{-1}\right| \Delta\right)^{-1}, \\ x^* &= M\left(|x^c| + \left|(A^c)^{-1}\right| \delta\right). \end{aligned}$$

The parametric version of the Bauer–Skeel method for solving parametric interval linear systems with affine-linear dependencies was proposed in [80]. It is based on the following theorem:

Theorem 5.11 (Hladík [80]) *Let $A(\boldsymbol{p})x = b(\boldsymbol{p})$ be a parametric interval linear system with affine-linear dependencies such that $A(p^c)$ is non-singular. Define*

$$M^* = \sum_{k=1}^{K} \left|A(p^c)^{-1}A^{(k)}\right| p_k^{\Delta},$$
$$x^c = A(p^c)^{-1}b(p^c)$$
$$\delta = \sum_{k=1}^{K} \left|A(p^c)^{-1}\left(b^{(k)} - A^{(k)}x^c\right)\right| p_k^{\Delta}.$$

If $\rho(M^*) < 1$, *then*

$$S(\boldsymbol{p}) \subseteq x^c + (I - M^*)^{-1}\delta[-1, 1]. \tag{5.25}$$

It can be seen from formula (5.25) that, in the affine-linear case, the method described by Theorem 5.11 is equivalent to the Direct Method with $R = A(p^c)^{-1}$. Indeed, we have $\boldsymbol{B} = I + B^{\Delta}[-1, 1]$, where $B^{\Delta} = \sum_{k=1}^{K} \left|A(p^c)^{-1}A^{(k)}\right| p_k^{\Delta}$ (see Theorem 3.34). Hence, $\langle \boldsymbol{B} \rangle = I - B^{\Delta}$. However, Theorem 5.11 does not exactly correspond to Theorem 5.10. If we follow the reasoning presented by Rohn [217], then we will obtain the following formulae for Parametric Bauer–Skeel (PBS) method.

Let x^c be the solution to midpoint system $A(p^c)x = b(p^c)$. If $A(p)x = b(p)$, for some $p \in \boldsymbol{p}$, then

$$\begin{aligned}
\left|x - x^c\right| &= \left|A(p^c)^{-1}A(p^c)\left(x - x^c\right)\right| = \\
&= \left|A(p^c)^{-1}\left(\left(A(p) - A(p^c\right)x + b(p) - b(p^c)\right)\right| = \\
&= \left|\sum_{k=1}^{K}\left(A(p^c)^{-1}A^{(k)}\right)\left(p_k^c - p_k\right)x + \sum_{k=1}^{K}\left(A(p^c)^{-1}b^{(k)}\right)\left(p_k - p_k^c\right)\right| \leqslant \\
&\leqslant \sum_{k=1}^{K}\left|A(p^c)^{-1}A^{(k)}\right| p_k^{\Delta}|x| + \sum_{k=1}^{K}\left|A(p^c)^{-1}b^{(k)}\right| p_k^{\Delta} = B^{\Delta}|x| + z^{\Delta},
\end{aligned}$$

where $\boldsymbol{B} = \text{hull}\left(\left\{A(p^c)^{-1}A(p) \mid p \in \boldsymbol{p}\right\}\right)$, $\boldsymbol{z} = \text{hull}\left(\left\{A(p^c)^{-1}b(p) \mid p \in \boldsymbol{p}\right\}\right)$. So

$$\begin{aligned}
\left|x - x^c\right| &\leqslant B^{\Delta}|x| + z^{\Delta} = B^{\Delta}|x - x^c + x^c| + z^{\Delta} \leqslant \\
&\leqslant B^{\Delta}|x - x^c| + B^{\Delta}|x^c| + z^{\Delta},
\end{aligned}$$

and hence

$$\left(I - B^{\Delta}\right)|x - x^c| \leqslant B^{\Delta}|x^c| + z^{\Delta}.$$

If $\rho(B^{\Delta}) < 1$, then $(I - B^{\Delta})^{-1} \geqslant 0$. Premultiplying by $M = (I - B^{\Delta})^{-1}$ gives

$$|x - x^c| \leqslant M\left(B^{\Delta}|x^c| + z^{\Delta}\right) = MB^{\Delta}|x^c| + Mz^{\Delta}.$$

Hence

$$\left|x - x^c\right| \leqslant (M - I)\left|x^c\right| + Mz^{\Delta} = M\left(\left|x^c\right| + z^{\Delta}\right) - \left|x^c\right| = x^* - \left|x^c\right|,$$

and we have

$$S(\mathbf{p}) \subseteq x^c + \left(x^* - |x^c|\right)[-1, 1]. \tag{5.26}$$

As shows numerical experiments, bounds given by (5.26) are rather crude. In order to improve them, we can employ the residual correction as described in Sect. 5.1.3. The PBS with residual correction (PBSRC) is mathematically equivalent to the method described by Theorem 5.11 and thus also to the DM method. The asymptotic time complexity of the PBSRC algorithm is $\mathcal{O}(Kn^3)$.

As in the case of DM, the PBSRC method can be quite straightforwardly extended to parametric interval linear systems with affine-linear dependencies and a multiple right-hand side.

Theorem 5.12 *Let $A(\mathbf{p})x = B(\mathbf{p})$ be a parametric interval linear system with a multiple right-hand side and suppose that $A(p^c)$ is non-singular. Define $C^\Delta = \sum_{k=1}^{K} \left|A(p^c)^{-1}A^{(k)}\right| p_k^\Delta$. If $\rho\left(C^\Delta\right) < 1$, then*

$$S(\mathbf{p}) \subseteq X^c + MZ^\Delta[-1, 1], \tag{5.27}$$

where

$$M = \left(I - C^\Delta\right)^{-1},$$
$$X^c = A(p^c)^{-1}B(p^c),$$
$$Z^\Delta = \sum_{k=1}^{K} \left|A(p^c)^{-1}\left(B^{(k)} - A^{(k)}X^c\right)\right| p_k^\Delta.$$

The next theorem generalizes Bauer–Skeel method with residual correction to parametric interval linear systems with arbitrary dependencies.

Theorem 5.13 *Let $A(\mathbf{p})x = b(\mathbf{p})$ be a parametric interval linear system, R be some matrix, and $\tilde{x}$ a real vector. Define $B(p) = RA(p)$, $z(p) = R\,(b(p) - A(p))$, and let $\mathbf{z}$ and $\mathbf{B}$ be the interval extensions of, respectively, $z(p)$ and $B(p)$ over $\mathbf{p}$ such that B^c is non-singular. If $\rho\left(\left|(B^c)^{-1}\right| B^\Delta\right) < 1$, then*

$$S(\mathbf{p}) \subseteq \tilde{x} + x^c + \left(x^* - |x^c|\right)[-1, 1], \tag{5.28}$$

where

$$x^c = \left(B^c\right)^{-1} z^c,$$
$$M = \left(I - \left|\left(B^c\right)^{-1}\right| B^\Delta\right)^{-1},$$
$$x^* = M\left(|x^c| + \left|\left(B^c\right)^{-1}\right| z^\Delta\right).$$

Proof If $x \in S(\boldsymbol{p})$, then $A(p)x = b(p)$ for some $p \in \boldsymbol{p}$. So, $u = x - \tilde{x}$ solves system

$$A(p)u = b(p) - A(p)\tilde{x}.$$

Since $RA(p) \in \boldsymbol{B}$ and $R\,(b(p) - A(p)\tilde{x}) \in \boldsymbol{z}$, hence u is a solution to interval linear system

$$\boldsymbol{B}u = \boldsymbol{z},$$

So, $x \in \tilde{x} + S(\boldsymbol{B}, \boldsymbol{z})$ and (5.28) follows directly from Theorem 5.10. □

5.1.5 Generalized Parametric Hansen–Bliek–Rohn Method

The non-parametric version of the *Hansen–Bliek–Rohn* (HBR) method is based on the following theorem (cf. [217]):

Theorem 5.14 (Hansen [75], Bliek [20], Rohn [209]) *Let $\boldsymbol{A}x = \boldsymbol{b}$ be an interval linear system. If spectral radius $\rho\left(\left|(A^c)^{-1}\right| A^{\Delta}\right) < 1$, then each $A \in \boldsymbol{A}$ is non-singular and*

$$S(\boldsymbol{A}, \boldsymbol{b}) \subseteq [\min\{x_d, Tx_d\}, \max\{x_u, Tx_u\}], \tag{5.29}$$

where

$$\begin{aligned}
M &= \left(I - \left|(A^c)^{-1}\right| \Delta\right)^{-1}, \\
D &= \operatorname{diag}\{M_{11}, \ldots, M_{nn}\}, \\
T &= (2D - I)^{-1}, \\
x^* &= M\left(|x^c| + \left|(A^c)^{-1}\right| \delta\right) \\
x_d &= -x^* + D(x^c + |x^c|), \\
x_u &= x^* + D(x^c - |x^c|).
\end{aligned}$$

Hladík generalized the HBR method to parametric interval linear systems with affine-linear dependencies according to the following theorem:

Theorem 5.15 (Hladík [80]) *Let $A(\boldsymbol{p})x = b(\boldsymbol{p})$ be a parametric interval linear system with affine-linear dependencies such that $A(p^c)$ is non-singular. Define M^* and x^c as in Theorem 5.11. If $\rho(M^*) < 1$, then*

$$S(\boldsymbol{p}) \subseteq [\min\{x_d, Tx_d\}, \max\{x_u, Tx_u\}], \tag{5.30}$$

where

$$
\begin{aligned}
M &= (I - M^*)^{-1}, \\
D &= \mathrm{diag}\{M_{11}, \dots, M_{nn}\}, \\
T &= (2D - I)^{-1}, \\
x^* &= M\left(|x^c| + \sum_{k=1}^{K} \left|A(p^c)^{-1}b^k\right| p_k^{\Delta}\right), \\
x_d &= -x^* + D(x^c + |x^c|), \\
x_u &= x^* + D(x^c - |x^c|).
\end{aligned}
$$

Rohn [217] proved (Theorem 3, [217]) that the Hansen–Bliek–Rohn bounds for the solution set of interval linear systems were never worse and "almost always" better than the Bauer–Skeel bounds. Whereas according to [80] the parametric versions of these methods are incomparable, i.e., sometimes one method is better and sometimes the other. The reason for this discrepancy is that in the non-parametric case both the BS method and the HBR method are applied to system $\boldsymbol{A}x = \boldsymbol{b}$ (without residual correction). Whereas, in [80], the PBS method is applied to system $A(\boldsymbol{p}) = b(\boldsymbol{p}) - A(\boldsymbol{p})x^c$ (with residual correction) and the PHBR method to system $A(\boldsymbol{p})x = b(\boldsymbol{p})$ (without residual correction).

Let us now compare the PBS and PHBR bounds. For the comparison purposes we will denote PBS bounds (5.26) by $\underline{x}$, $\overline{x}$, and PHBR bounds (5.30) by $\underline{\underline{x}}$, $\overline{\overline{x}}$. First, we will prove that $M_{ii} \geqslant 1$ for each i. If $\rho\left(B^{\Delta}\right) < 1$, then $M = (I - B^{\Delta})^{-1} \geqslant 0$ and

$$
M - I = \sum_{i=0}^{\infty}\left(B^{\Delta}\right) - I = \sum_{i=1}^{\infty}\left(B^{\Delta}\right) = MB^{\Delta} \geqslant 0,
$$

which means that $M_{ii} \geqslant 1$. The next theorem applies straightforwardly to parametric case.

Theorem 5.16 (Rohn [217]) *Under the common assumption* $\rho(B^{\Delta}) < 1$, *for each* $i = 1, \dots, n$, *we have*

$$
\overline{x}_i - \overline{\overline{x}}_i \geqslant \min\left\{(M_{ii} - 1)(|x_i^c| - x_i^c), \frac{2(M_{ii} - 1)}{2M_{ii} - 1}(x_i^* - |x_i^c|)\right\} \geqslant 0, \qquad (5.31\text{a})
$$

$$
\underline{\underline{x}}_i - \underline{x}_i \geqslant \min\left\{(M_{ii} - 1)(|x_i^c| + x_i^c), \frac{2(M_{ii} - 1)}{2M_{ii} - 1}(x_i^* - |x_i^c|)\right\} \geqslant 0. \qquad (5.31\text{b})
$$

Using a simple example from [80] we will show that the residual correction of the right-hand vector has significant impact on the accuracy of the PBS results.

Table 5.1 Comparison of results obtained using DM, PBS, PBSRC and PHBR methods for Example 5.17

DM	PBSRC [80]	PHBR [80]	PBS	PHBR
[0.1282, 1.2051]	[0.1282, 1.2051]	[−0.4359, 3.7692]	−2.4359, 3.7692]	[−0.4359, 3.7692]
[−1.4103, − 0.3675]	[−1.4103, − 0.3675]	[−4.8718, − 0.0923]	[−4.8718, 3.094]	[−4.8718, − 0.0923]

Example 5.17 Consider the following two-dimensional system:

$$\begin{pmatrix} p_1 & p_2 - 1 \\ p_2 & p_1 \end{pmatrix} \begin{pmatrix} x_1 \\ x_2 \end{pmatrix} = \begin{pmatrix} -p_1 + \frac{1}{3} \\ p_2 \end{pmatrix},$$

where $p_1 \in [-2, -1]$ and $p_2 \in [3, 5]$. Table 5.1 presents the results of the PBS and PHBR methods together with the results obtained in [80] and the result of the DM method. Additionally, all the obtained enclosures and the parametric solution set are depicted in Fig. 5.2.

It can be seen from Table 5.1 that the difference between the PBS and PHBR bounds is

$$\begin{aligned} \overline{x}_1 - \overline{\overline{x}}_1 &= 0, & \underline{\underline{x}}_1 - \underline{x}_1 &\approx 2, \\ \overline{x}_2 - \overline{\overline{x}}_2 &\approx 3.19, & \underline{\underline{x}}_2 - \underline{x}_2 &= 0, \end{aligned}$$

which, as can be easily checked, agrees with formula (5.31a). It can be seen as well that the PBSRC bounds are equal to the DM bounds and they are much better than PBS and PHBR bounds. The residual correction turned out to be crucial for the quality of the PBS bounds. So, let us now consider the PHBR method with residual correction (PHBRC).

Corollary 5.18 *Let $A(\boldsymbol{p})x = b(\boldsymbol{p})$ be a parametric interval linear system with affine-linear dependencies such that $A(p^c)$ is non-singular. Define M^*, x^c and δ as in Theorem 5.11. If $\rho(M^*) < 1$, then*

$$S(\boldsymbol{p}) \subseteq x^c + (I - M^*)^{-1}\delta[-1, 1]. \tag{5.32}$$

Proof If x is a solution to system $A(\boldsymbol{p})x = b(\boldsymbol{p})$, then $u = x - x^c$ is a solution to system $A(\boldsymbol{p})u = b(\boldsymbol{p}) - A(\boldsymbol{p})x^c$. Now, it is enough to observe that the midpoint solution to the new system is the zero vector and (5.32) follows directly from Theorem 5.15. □

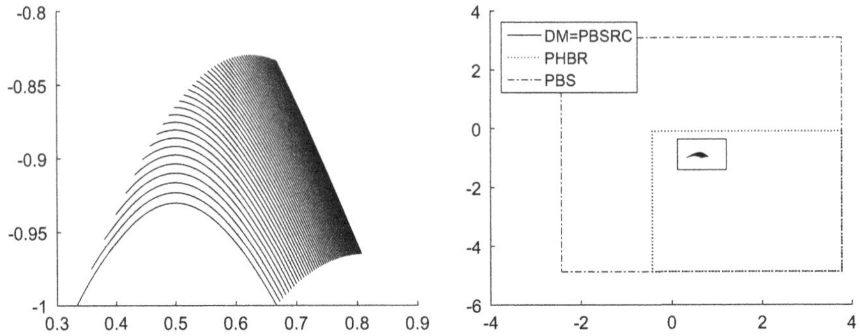

Fig. 5.2 Solution set of system from Example 5.17 (left); solution set (black region) together with DM, PBS, PBSRC and PHBR bounds (right)

Table 5.2 Comparison of PHBR and PHBRC enclosures for Example 5.19

PHBR	PHBRC
[0.7272727272726,1.333333333333577]	[0.6666666666664821,1.333333333333575]
[0.9999999999997984,1.000000000000145]	[0.9999999999998003,1.000000000000143]

From (5.32) it follows that the PHBRC method is equivalent to both DM and PBSRC methods.

The residual correction usually improves the PHBR bounds. However, as shows the next example, this is not always the case.

Example 5.19 Consider parametric interval linear system [185]

$$\begin{pmatrix} p_1 & p_1 \\ p_1 & p_1 + 0.01 \end{pmatrix} \cdot \begin{pmatrix} x_1 \\ x_2 \end{pmatrix} = \begin{pmatrix} p_2 \\ p_2 + 0.01 \end{pmatrix},$$

where $p_1 \in [0.9, 1.1]$, $p_2 \in [1.9, 2.1]$. Table 5.2 shows the enclosures obtained using the PHBR and PHBRC method.

The above observations suggest that the best idea is to combine the PHBR and PHBRC to get better bounds. In what follows, the PHBRC will refer to the combined PHBR method. The sketch of the PHBRC method is presented in Algorithm 7. The time complexity of this algorithm is $\mathcal{O}(Kn^3)$, however, the method is obviously more time consuming that then its basic version.

The next theorem generalizes the PHBR method with residual correction to parametric interval linear systems with arbitrary dependencies.

Theorem 5.20 *Let $A(\boldsymbol{p})x = b(\boldsymbol{p})$ be a parametric interval linear system, R be some matrix, and $\tilde{x}$ a real vector. Define $B(p) = RA(p)$, $z(p) = R(b(p) - A(p)\tilde{x}$, and let $\boldsymbol{z}$ and $\boldsymbol{B}$ be the interval extensions of, respectively, $z(p)$ and $B(p)$ over $\boldsymbol{p}$ such that B^c is non-singular. If $\rho\left(\left|(B^c)^{-1}\right| B^{\Delta}\right) < 1$, then*

$$S(\boldsymbol{p}) \subseteq \tilde{x} + [\min\{x_d, Tx_d\}, \max\{x_u, Tx_u\}], \tag{5.33}$$

where

$$\begin{aligned}
x^c &= \left(B^c\right)^{-1} z^c, \\
M &= \left(I - \left|\left(B^c\right)^{-1}\right| B^{\Delta}\right)^{-1}, \\
D &= \mathrm{diag}(M_{11}, \ldots, M_{nn}), \\
T &= (2D - I)^{-1}, \\
x^* &= M\left(\left|x^c\right| + \left|\left(B^c\right)^{-1}\right| z^{\Delta}\right), \\
x_d &= -x^* + D\left(x^c + \left|x^c\right|\right), \\
x_u &= x^* + D(x^c - \left|x^c\right|).
\end{aligned}$$

Proof Analogous to the proof of Theorem 5.13. □

Algorithm 7 Parametric Hansen–Bliek–Rohn (PHBR) method

Input: $A(\boldsymbol{p})x = b(\boldsymbol{p})$, $\boldsymbol{p}$
Output: $\boldsymbol{x}^{OI}$
  $R \approx A(p^c)^{-1}$
  $x^c = Rb(p^c)$
  $B(p) = RA(p)$
  $z(p) = Rb(p)$
  $w(p) = R(b(p) - A(p)x^c)$
  $\boldsymbol{B} = \mathrm{hull}\left(\{B(p) \mid p \in \boldsymbol{p}\}\right)$
  $\boldsymbol{z} = \mathrm{hull}\left(\{z(p) \mid p \in \boldsymbol{p}\}\right)$
  $\boldsymbol{w} = \mathrm{hull}\left(\{w(p) \mid p \in \boldsymbol{p}\}\right)$
  if $(\rho\left(B^{\Delta}\right) < 1)$ **then**
    $M = (I - B^{\Delta})^{-1}$
    $D = \mathrm{diag}(M_{11}, \ldots, M_{nn})$
    $T = (2D - I)^{-1}$
    $x^* = M(|x^c| + z^{\Delta})$
    $x^{**} = Mw^{\Delta}$
    $x_d = -x^* + D(x^c + |x^c|)$
    $x_u = x^* + D(x^c - |x^c|)$
    $\boldsymbol{x}^{OI} = [\max\{\min\{x_d, Tx_d\}, x^c - x^{**}\}, \min\{\max\{x_u, Tx_u\}, x^c + x^{**}\}]$
    return "Success"
  else
    return "Failure"
  end if

The PHBR method can be extended for solving parametric interval linear systems with affine-linear dependencies and a multiple right-hand side in a similar manner as the DM and the PHBR methods (cf. [37]).

5.1.6 Monotonicity Approach

The approach presented in this section (cf. Skalna [250]) exploits the monotonicity properties of real valued functions. If x is a solution to a parametric interval linear system, then there is $p \in \boldsymbol{p}$ such that $A(p)x = b(p)$. If $A(p)$ is non-singular, then $x = A(p)^{-1}b(p)$, which means that, for $i = 1, \ldots, n$, x_i is a real valued function of p, i.e., $x_i = x_i(p)$. According to the considerations presented in Sect. 1.3, if $x_i(p)$ is monotonic on $\boldsymbol{p}$ with respect to all parameters, then $\text{Rge}\,(x_i \mid \boldsymbol{p})$ is spanned by the values of x_i at the respective endpoints of $\boldsymbol{p}$. If x_i is monotonic with respect to only some parameters, we can then fix these parameters at the respective endpoints and then bound the range of x_i on a box of a lower dimension according to Theorem 1.15.

The global monotonicity property of x_i with respect to p_k can be verified by determining the sign of partial derivative $\frac{\partial x_i}{\partial p_k}(p)$ on $\boldsymbol{p}$. So, assume that x_i, A_{ij} and b_i $(i,\ j = 1,\ \ldots,\ n)$ are continuously differentiable in $\boldsymbol{p}$. The partial derivatives of x with respect to p_k satisfy the following system of parametric equations:

$$A(p)\frac{\partial x}{\partial p_k}(p) = \frac{\partial b}{\partial p_k}(p) - \frac{\partial A}{\partial p_k}(p)x(p). \tag{5.34}$$

In order to bound the ranges of $\frac{\partial x_i}{\partial p_k}(p)$ on $\boldsymbol{p}$, we must solve parametric interval linear system

$$\left\{A(p)\frac{\partial x}{\partial p_k}(p) = \frac{\partial b}{\partial p_k}(p) - \frac{\partial A}{\partial p_k}(p)x(p),\ \ p \in \boldsymbol{p}\right\}. \tag{5.35}$$

However, since $x(\boldsymbol{p})$ is not known, it must be replaced by an outer interval enclosure $\boldsymbol{x}^{OI}$. This obviously causes some lost of information since $\boldsymbol{x}^{OI}$ does not retain information about the dependencies, i.e., the elements of $\boldsymbol{x}^{OI}$ are treated as new interval parameters that are independent of $p_1, \ldots, p_K$. This gives the following parametric interval linear system:

$$\left\{A(p)\frac{\partial x}{\partial p_k}(p, q) = \frac{\partial b}{\partial p_k}(p) - \frac{\partial A}{\partial p_k}(p)q,\ q \in \boldsymbol{x}^{OI},\ p \in \boldsymbol{p}\right\}. \tag{5.36}$$

For the affine-linear case, formula (5.36) takes on a simpler form:

$$\left\{ A(p)\frac{\partial x}{\partial p_k}(p,q) = b^{(k)} - A^{(k)}q,\ q \in \boldsymbol{x}^{OI}, p \in \boldsymbol{p} \right\}, \tag{5.37}$$

and moreover, no additional tools (e.g., automatic differentiation) are needed.

Let $\boldsymbol{d}^k$ denote the solution of system (5.36). By solving Eq. (5.36) for $k = 1, \ldots, K$, we obtain vectors $\boldsymbol{d}^1, \ldots, \boldsymbol{d}^K$. Now, assume that, for each $k \in \{1, \ldots, K\}$ and for each $i \in \{1, \ldots, n\}$, it holds that $\boldsymbol{d}_i^k \geqslant 0$ or $\boldsymbol{d}_i^k \leqslant 0$. In order to obtain the bounds on x_i, we put, for $k = 1, \ldots, K$,

$$\underline{p}_k^i = \begin{cases} \underline{p}_k, & \boldsymbol{d}_i^k \geqslant 0, \\ \overline{p}_k, & \boldsymbol{d}_i^k \leqslant 0, \end{cases} \qquad \overline{p}_k^i = \begin{cases} \overline{p}_k, & \boldsymbol{d}_i^k \geqslant 0, \\ \underline{p}_k, & \boldsymbol{d}_i^k \leqslant 0, \end{cases} \tag{5.38}$$

and solve a pair of real linear systems

$$A(\underline{p}^i)u^i = b(\underline{p}^i), \tag{5.39a}$$

$$A(\overline{p}^i)v^i = b(\overline{p}^i). \tag{5.39b}$$

Then, Rge $(x_i \mid \boldsymbol{p}) = [u_i^i, v_i^i]$. By solving n pairs of real linear systems (5.39), we obtain range Rge $(x \mid \boldsymbol{p})$. The number of equations to be solved can be decreased by removing the redundant vectors from list

$$L_1 = \{\underline{p}^1, \ldots, \underline{p}^n, \overline{p}^1, \ldots, \overline{p}^n\}.$$

If only some of the partial derivatives have a constant sign on $\boldsymbol{p}$, then instead of $2n$ real systems, we must solve $2n$ parametric interval linear systems in the worst case, however with a smaller number of interval parameters. Since the widths of the resulting intervals usually depend on the number of parameters, hence narrower bounds should be generally obtained.

The main deficiency of the above-described approach follows from replacing $x(\boldsymbol{p})$ by $\boldsymbol{x}^{OI}$ in (5.35). This can be overcome by using the following modification (which was suggested by Kolev). For each $k = 1, \ldots, K$, create the following parametric interval linear system:

$$\left\{ \begin{pmatrix} A(p) & 0 \\ \frac{\partial A}{\partial p_k}(p) & A(p) \end{pmatrix} \begin{pmatrix} x(p) \\ \frac{\partial x}{\partial p_k}(p) \end{pmatrix} = \begin{pmatrix} b(p) \\ \frac{\partial b}{\partial p_k}(p) \end{pmatrix},\ p \in \boldsymbol{p} \right\}, \tag{5.40}$$

and solve it in order to obtain bounds (hopefully narrower) for $\frac{\partial x}{\partial p_k}(p)$ over $\boldsymbol{p}$. Then, we proceed analogously as described above. Notice that, for the affine-linear case, system (5.41) takes the following form:

$$\left\{ \begin{pmatrix} A(p) & 0 \\ A^{(k)} & A(p) \end{pmatrix} \begin{pmatrix} x(p) \\ \frac{\partial x}{\partial p_k}(p) \end{pmatrix} = \begin{pmatrix} b(p) \\ b^{(k)} \end{pmatrix},\ p \in \boldsymbol{p} \right\}. \tag{5.41}$$

The presented modified version of the monotonicity approach eliminates the problem of the "lost of information", however, we instead must solve a number of systems that are twice as large as the original system. Since the basic version of the method is already quite expensive, the modified method might be inefficient for large problems.

Below, we present a new approach to the "lost of information" problem that relies on replacing $x(p)$ in the right-hand side of (5.34) with p-solution $\boldsymbol{x}(p) = Lp + \boldsymbol{a}$ (see Sect. 5.2). As a result, we obtain the following parametric interval linear system:

$$\left\{ A(p)\frac{\partial x}{\partial p_k}(p) = b(p, a) - C(p) - \frac{\partial A}{\partial p_k}(p)a,\ p \in \boldsymbol{p}, a \in \boldsymbol{a} \right\}, \tag{5.42}$$

where $C(p) = \frac{\partial A}{\partial p_k}(p)Lp$, and the elements of interval vector $\boldsymbol{a}$ are treated as new interval parameters, independent of $\boldsymbol{p}$. In the affine-linear case, linear system (5.42) takes the following form:

$$\left\{ A(p)\frac{\partial x}{\partial p_k}(p) = b^{(k)} - C(p) - A^{(k)}a,\ p \in \boldsymbol{p}, a \in \boldsymbol{a} \right\}, \tag{5.43}$$

where $C(\boldsymbol{p}) = \{A^{(k)}Lp \mid p \in \boldsymbol{p}\}$.

For further analysis, we will refer to the basic version of the monotonicity approach as the MA method, the monotonicity approach involving the system (5.35) will be referred to as the MA1 method, and the third approach will be referred to as the MA2 method. In the affine-linear case, the asymptotic time complexity of the MA and MA1 methods is $\mathcal{O}(Kn^2 + K\kappa)$, where κ is the complexity of the method used to solve parametric interval linear systems (5.37) and (5.41), respectively. The asymptotic time complexity of the MA2 method is $\mathcal{O}(Kn^2 + K\kappa + \tau)$, where τ is the time complexity of the method used to compute the p-solution. In the general case, the time complexity will depend on the method used for range bounding. The computational time of all three methods can be decreased by performing the computation in parallel. The sketch of the algorithm of the Monotonicity Approach employing the p-solution is presented in Algorithm 8.

Algorithm 8 Monotonicity Approach employing p-solution

Input: $A(\boldsymbol{p})x = b(\boldsymbol{p})$, $\boldsymbol{p}$
Output: $\boldsymbol{x}^{OI}$
if ($A(\boldsymbol{p})$ is strongly regular) **then**
  Compute the p-solution $\boldsymbol{x}(p) = Lp + \boldsymbol{a}$
  $\boldsymbol{p}' = (\boldsymbol{p}_1, \ldots, \boldsymbol{p}_K, \boldsymbol{a}_1, \ldots, \boldsymbol{a}_n)$
  Create $A'(p')$
  for k=1 **to** K **do**
    Create $b'(p')$
    Solve system $A'(\boldsymbol{p}')\frac{\partial x}{\partial p_k} = b'(\boldsymbol{p}')$
    for $i = 1$ **to** n **do**
      if ($\frac{\partial x}{\partial p_k} > 0$) **then**
        $\underline{p}^i_k = \underline{p}'_k; \overline{p}^i_k = \overline{p}'_k$
      else if ($\frac{\partial x}{\partial p_k} < 0$) **then**
        $\underline{p}^i_k = \overline{p}'_k; \underline{p}^i_k = \overline{p}'_k$
      else
        $\underline{p}^i_k = p'_k; \overline{p}^i_k = p'_k$
      end if
    end for
  end for
  for $i = 1$ **to** n **do**
    Solve system $A(\underline{p}^i)u = b(\underline{p}^i)$
    Solve system $A(\overline{p}^i)v = b(\overline{p}^i)$
    $\boldsymbol{x}^{OI}_i = [u_i, v_i]$
  end for
  return "Success"
else
  return "Failure"
end if

5.2 Methods for Computing Parametric Solution

In this section, we discuss the methods for computing the parametric solution (p-solution) to parametric interval linear systems [113]. We start from a quick overview of the methods for computing the p-solution of a parametric interval linear system with affine-linear dependencies [113, 115, 117]. Then, we present the results of our own research.

5.2.1 Parametric Methods

The concept of the p-solution was introduced by Kolev in [113]. Therein, a *Parametric Linear Iterative Method* (PLIM) for computing the p-solution of parametric interval linear systems with affine-linear dependencies was also proposed. The iterative scheme of the PLIM method is given by

$$v^{(0)} = 0, \tag{5.44a}$$

$$v^{i+1}(p) = B(p)v^{(k)}(p) + C^{(0)}p, \tag{5.44b}$$

where

$$\begin{aligned} B(p) &= -\left(\sum_{k=1}^{K} \left(A^{(0)}\right)^{-1} A^{(k)} p_k\right), \\ C^{(0)} &= \left(C_{*1}^{(0)} \dots C_{*K}^{(0)}\right) \\ C_{*k}^{(0)} &= \left(A^{(0)}\right)^{-1} \left(b^{(k)} - A^{(k)} x^{(0)}\right), \ k = 1, \dots, K, \end{aligned}$$

C_{*k} denotes the kth column of matrix C, $x^{(0)}$ is a solution to midpoint system $A^{(0)}x = b^{(0)}$, and $v = x - x^{(0)}$. The method assumes that the parameters vary within the intervals of $[-1, 1]$, but this is not a limitation since a parametric interval linear system with affine linear dependencies can be normalized using the affine transformation from Sect. 3.3. The method is called linear since the quadratic forms that arise in each iteration due to the multiplication of $B(p)$ and $v^i(p)$ are approximated outwardly by linear interval (LI) forms using Procedure 1 (see [113]).

The *Parametric Quadratic Iterative Method* (PQIM) for computing the p-solution proposed by Kolev in [117] is based on iterative scheme (5.44), however, instead of a linear interval form, a quadratic interval (QI) form is used to outwardly approximate higher order terms. Despite the use of higher-order approximations (which preserve more information about dependencies), the quadratic iterative method sometimes gives worse results than the linear iterative method. The derivation of both methods is a bit tedious, therefore, it is omitted here. For details, we refer the readers to the source [113, 117].

The *Parametric Direct Method* (PDM) for computing the parametric solution was proposed by Kolev in [115]. The general idea of the basic version of this method is similar to that of the Direct Method from Sect. 5.1.3. First, the residual correction of the right-hand vector is performed using solution $x^{(0)}$ to mid-point system $A^{(0)}x = b^{(0)}$ (also in this case it is assumed that all parameters vary within interval $[-1, 1]$). Then, the system is pre-conditioned with the inverse of the mid-point matrix. The obtained parametric interval linear system

$$B(\boldsymbol{p})v = b(\boldsymbol{p}), \tag{5.45}$$

where $v = x - x^{(0)}$, is transformed to the so-called "mixed type" system, with an interval matrix and the parametric right-hand vector:

$$\boldsymbol{B}v = b(\boldsymbol{p}), \tag{5.46}$$

where $b^{(0)} = 0$ and $\boldsymbol{B}$ is such that $B^c = I$ and $B^{\Delta} = \sum_{k=1}^{K} \left|B^{(k)}\right|$. Partial ignorance of the dependencies in (5.46) causes some loss of information; however, the computation is much easier. Under the assumption that $\rho(B^{\Delta}) < 1$ (i.e., $\boldsymbol{B}$ is strongly regular), the inverse of matrix $\boldsymbol{B}$ is given by (Theorem 4.4, [219])

$$\boldsymbol{H} = \boldsymbol{B}^{-1} = [-M + \mathrm{diag}(z), M], \tag{5.47}$$

where

$$M = \left(I - B^{\Delta}\right)^{-1} \geqslant 0,$$
$$z_i = \frac{2M_{ii}^2}{2M_{ii} - 1}, \quad i = 1, \ldots, n.$$

Then, the p-solution obtained using the Parametric Direct Method is given by

$$\boldsymbol{x}(p) = Lp + \boldsymbol{a}, \tag{5.48}$$

where

$$L = (L_{*1} \ldots L_{*K}) \in \mathbb{R}^{n \times K}, \tag{5.49a}$$
$$L_{*k} = H^c b^{(k)}, \tag{5.49b}$$
$$\boldsymbol{a} = x^{(0)} + \sum_{k=1}^{K} H^{\Delta} \left|b^{(k)}\right| [-1, 1]. \tag{5.49c}$$

Theorem 5.21 ([115]) *Let $A(\boldsymbol{p})x = b(\boldsymbol{p})$ be a parametric interval linear system with affine-linear dependencies such that $A^{(0)}$ is non-singular and assume that $\rho(B^{\Delta}) < 1$, where $\boldsymbol{B} = \mathrm{hull}\left(\{(A^{(0)})^{-1}A(p) \mid p \in \boldsymbol{p}\}\right)$. Then*

- *$A(\boldsymbol{p})$ is strongly regular,*
- *the p-solution $\boldsymbol{x}(p)$ of parametric interval linear systems system $A(\boldsymbol{p})x = b(\boldsymbol{p})$ exists and is determined by (5.48), (5.49).*

Kolev [115] proved that the Parametric Direct Method and the Direct Method from Sect. 5.1.3 yield the same interval enclosures.

Corollary 5.22 ([115]) *Let $\boldsymbol{x}(p)$ be given by (5.48), (5.49). Then* hull $(\{\boldsymbol{x}(p) \mid\})$ *$\{p \in \boldsymbol{p}\}$ equals to the outer interval solution produced by the Direct Method [248].*

Table 5.3 Comparison of parametric methods for Example 5.23

PLIM	PQIM	PDM	DM
$[-0.151, 0.744]$	$[-0.156, 0.750]$	$[-0.291, 0.862]$	$[-0.291, 0.862]$
$[-0.054, 0.141]$	$[-0.054, 0.140]$	$[-0.089, 0.185]$	$[-0.089, 0.185]$
$[-2.350, -0.810]$	$[-2.367, -0.796]$	$[-2.501, -0.642]$	$[-2.501, -0.642]$

Example 5.23 Consider the following parametric interval linear system with affine-linear dependencies (cf. [113]):

$$\begin{pmatrix} p_1 & p_2+1 & -p_3 \\ p_2+1 & -3 & p_1 \\ 2-p_3 & 4p_2+1 & 1 \end{pmatrix} \begin{pmatrix} x_1 \\ x_2 \\ x_3 \end{pmatrix} = \begin{pmatrix} 2p_1 \\ p_3-1 \\ -1 \end{pmatrix},$$

where $p \in [1-\delta, 1+\delta]$. The results for $\delta = 0.3$ are presented in Table 5.3.

The results of the above described parametric methods and the DM method are presented in Table 5.3.

It can be seen that the PDM and DM methods were significantly outperformed by both the PLIM and PQIM methods. Moreover, the Linear Iterative Method produced better bounds for the first and third component of the solution set than the Parametric Quadratic Iterative Method.

5.2.2 *Interval-Affine Gauss–Seidel Iteration*

The parametric methods described in the previous section are very fast, since the computation is performed on floating-point numbers. However, the produced p-solutions might not yield guaranteed bounds for the parametric solution set. In (Skalna and Hladík [259]), we proposed the *Interval-affine Gauss–Seidel Iteration* (IAGSI) method for computing the p-solution of parametric interval linear systems, which provides guaranteed bounds, however, at the expense of higher computational time. The general idea of this method is as follows. We pre-condition the system with the inverse of the mid-point matrix, $\left(C^{(0)}\right)^{-1}$, which is an optimal pre-conditioner from a certain point of view (cf. [83]) [255]. Pre-conditioning transforms system (4.25) into a new system

$$V(e)x = v(e), \tag{5.50}$$

where $V(e) = \left(C^{(0)}\right)^{-1} C(e)$ and $v(e) = \left(C^{(0)}\right)^{-1} c(e)$. The solution set of system (5.50) is larger than the solution set of system (4.25), however, system (5.50) usually has better convergence properties.

Proposition 5.24 *The solution set of the system (4.25) is included in the solution set of the system (5.50).*

Now, the p-solution is computed in the following way. First, a rigorous enclosure $\boldsymbol{x}^0 \in \mathbb{IR}^n$ for the solution set of system (5.50) is computed. Elements $\boldsymbol{x}_i^0$ $(i = 1, \ldots, n)$ of $\boldsymbol{x}^0$ are then replaced by the revised affine forms: $\hat{x}_i^0 = (x_i^0)^c + (x_i^0)^\Delta[-1, 1]$. Finally, an iterative method is used to improve the initial enclosure. Here, the Gauss–Seidel Iteration method is used, which is one of the best-known stationary iterative methods for solving linear systems, however, it can be replaced by any other iterative method.

For pre-conditioned system (5.50), the Interval-affine Gauss–Seidel Iteration takes the following form:

$$\boldsymbol{x}^0(e) = \hat{\boldsymbol{x}}^0(e), \tag{5.51a}$$

$$\boldsymbol{x}^{i+1}(e) = (\boldsymbol{D}(e) + \boldsymbol{L}(e))^{-1}\left(\boldsymbol{v}(e) - \boldsymbol{U}(e)\boldsymbol{x}^i(e)\right), \tag{5.51b}$$

where the components of $\boldsymbol{D}(e)$, $\boldsymbol{L}(e)$, and $\boldsymbol{U}(e)$ are defined by

$$\boldsymbol{L}_{ij}(e) = \begin{cases} \boldsymbol{V}_{ij}(e), & i > j, \\ 0, & i \leqslant j, \end{cases} \quad \boldsymbol{U}_{ij}(e) = \begin{cases} \boldsymbol{V}_{ij}(e), & i < j, \\ 0, & i \geqslant j, \end{cases}$$

$$\boldsymbol{D}_{ij}(e) = \begin{cases} \boldsymbol{V}_{ii}(e), & i = j, \\ 0, & i \neq j. \end{cases}$$

The element-based formula can be obtained by taking advantage of the lower triangular form of $\boldsymbol{L}(e)$. Thus, for $j = 1, \ldots, n$,

$$\boldsymbol{x}_j^0(e) = \hat{\boldsymbol{x}}_j^0, \tag{5.52a}$$

$$\boldsymbol{x}_j^{i+1}(e) = \frac{\boldsymbol{v}_j(e) - \sum_{\mu=1}^{j-1} \boldsymbol{V}_{j\mu}(e)\boldsymbol{x}_\mu^{i+1}(e) - \sum_{\mu=j+1}^{n} \boldsymbol{V}_{j\mu}(e)\boldsymbol{x}_\mu^{i}(e)}{\boldsymbol{V}_{jj}(e)}. \tag{5.52b}$$

Clearly, for each $i \geqslant 1$, $\boldsymbol{x}^i(e)$ is a vector of revised affine forms (revised affine vector), the range of which is interval vector $\boldsymbol{x}^i = ([\boldsymbol{x}_1^{i+1}(e)], \ldots, [\boldsymbol{x}_n^{i+1}(e)])$. The stopping criterion of the IAGSI method employs distance measure (5.17).

Proposition 5.25 (Skalna and Hladík [259]) *For any $\boldsymbol{x}^i(e)$, $i \geqslant 1$, defined by (5.52), we have:*

(i) $[\boldsymbol{x}^i(e)]$ is an outer interval solution to systems (5.50) and (4.10),
(ii) revised affine vector $\boldsymbol{x}^i(e)$ determines a p-solution to systems (5.50) and (4.10).

Proof (i) If $\tilde{x}$ is a solution to the system (5.50), then there exist $V(e) \in \boldsymbol{V}(e)$ and $v(e) \in \boldsymbol{v}(e)$, such that

$$V(e)\tilde{x} = v(e). \tag{5.53}$$

Assuming that $V_{jj}(e) \neq 0$, the Eq. (5.53) can be written as

$$\tilde{x}_j = \frac{1}{V_{jj}(e)}\left(v_j(e) - \sum_{\mu\neq j} V_{j\mu}(e)\tilde{x}_\mu\right) \in$$
$$\in \frac{1}{\boldsymbol{V}_{jj}(e)}\left(\boldsymbol{v}_j(e) - \sum_{\mu\neq j} \boldsymbol{V}_{j\mu}(e)\hat{x}^0_\mu\right) \equiv \boldsymbol{x}'_j(e). \tag{5.54}$$

The interval vector $[\boldsymbol{x}'_j(e)]$ is a new outer interval enclosure for $\tilde{x}$. Since, in the jth step, an enclosure for $\tilde{x}_j$ and new enclosures for $\tilde{x}_1, \ldots, \tilde{x}_{j-1}$ are already available, they can be used in the right-hand side of (5.54). This gives

$$\tilde{x}_j \in \frac{1}{\boldsymbol{V}_{jj}(e)}\left(\boldsymbol{v}_j(e) - \sum_{\mu<j} \boldsymbol{V}_{j\mu}(e)\boldsymbol{x}_\mu(e) - \sum_{\mu>j} \boldsymbol{V}_{j\mu}(e)\hat{x}^0_\mu\right) = \boldsymbol{x}^1_j(e). \tag{5.55}$$

This means that for each $i \geqslant 1$, the interval vector $[\boldsymbol{x}^i(e)]$ is an outer interval enclosure for the solution set of the system (5.50). On account of Propositions 4.13 and 5.24, $[\boldsymbol{x}^i(e)]$ is also an outer interval enclosure for the solution set of the system (4.11).
(ii) Follows directly from the fact that the range of $\boldsymbol{x}^i(e)$ yields an outer interval solution to (4.11). □

Each revised affine vector $\boldsymbol{x}^i(e)$, $i \geqslant 1$, can be written as

$$\boldsymbol{x}^i(e) = \sum_{k=1}^{K} (x^i)^{(k)}\varepsilon_k + \boldsymbol{x}^i, \tag{5.56}$$

where $(x^i)^{(k)} \in \mathbb{R}^n$, $\boldsymbol{x}^i \in \mathbb{IR}^n$, or, equivalently, as

$$\boldsymbol{x}^i(e) = L^i e + \boldsymbol{x}^i, \tag{5.57}$$

where $L^i \in \mathbb{R}^{n\times K}$ with vectors $(x^i)^{(k)}$ as columns. Now, let

$$\lambda^i_j(e) = L^i_j e,\ e \in \boldsymbol{e}^K,\quad j = 1, \ldots, n, \tag{5.58}$$

where $L^i_j \in \mathbb{R}^K$ is the jth row of matrix L^i, and denote

$$\lambda^{i,\min}_j = \min\{\lambda^i_j(e), e \in \boldsymbol{e}^K\} = -\sum_{k=1}^{K} |(x^i)^{(k)}_j|,\quad j = 1, \ldots, n, \tag{5.59a}$$

$$\lambda^{i,\max}_j = \max\{\lambda^i_j(e), e \in \boldsymbol{e}^K\} = \sum_{k=1}^{K} |(x^i)^{(k)}_j|,\quad j = 1, \ldots, n. \tag{5.59b}$$

Theorem 5.26 (Skalna and Hladík [259]) *Let* $\boldsymbol{x}^i(e) = \sum_{k=1}^{K}(x^i)^{(k)}\varepsilon_k + \boldsymbol{x}^i$, *for* $i \geqslant 1$, *be the revised affine vector defined by (5.52), and let* $\lambda_j^{i,\min}$, $\lambda_j^{i,\max}$ *be given by (5.59). Then,* $\boldsymbol{\xi}$ *such that*

$$\boldsymbol{\xi}_j = \begin{cases} [\lambda_j^{i,\min} + \overline{x}_j^i, \lambda_i^{i,\max} + \underline{x}_j^i], & \lambda_j^{i,\min} + \overline{x}_j^i \leqslant \lambda_j^{i,\max} + \underline{x}_j^i, \\ \varnothing, & \text{otherwise}, \end{cases} \tag{5.60}$$

is an inner estimation of the hull solution to system (4.10) and system (5.50).

Proof If $\tilde{x}$ is a solution to the system (4.11), then there exists $p \in \boldsymbol{p}$, such that $\tilde{x} = A(p)^{-1}b(p)$. So, $\tilde{x}$ is a function of p, i.e., $\tilde{x} = \tilde{x}(p)$. Clearly, p can be written as $p = p^c e + p^\Delta$, and hence, for $j = 1, \ldots, n$, $\tilde{x}_j$ is a function of e, i.e., $\tilde{x}_j = f_j(e)$. From the construction of the iteration (5.52) it follows that, for $i = 1, \ldots, n$,

$$f_j(e) \in \lambda_j^i(e) + \boldsymbol{x}_j^i, \tag{5.61}$$

and hence

$$\lambda_j^i(e) + \underline{x}_j^i \leqslant f_j(e) \leqslant \lambda_j^i(e) + \overline{x}_j^i. \tag{5.62}$$

Denote the interval hull of the solution set of the system (4.11) as

$$\boldsymbol{x}^H = \left([x_1^{\min}, x_1^{\max}], \ldots, [x_n^{\min}, x_n^{\max}]\right), \tag{5.63}$$

where

$$x_j^{\min} = \min\{\tilde{x}_j(p) \mid p \in \boldsymbol{p}\}, \tag{5.64a}$$

$$x_j^{\max} = \max\{\tilde{x}_j(p) \mid p \in \boldsymbol{p}\}. \tag{5.64b}$$

On account of (5.61), the following inequalities hold true:

$$x_j^{\min} = \min\{f_j(e) \mid e \in \boldsymbol{e}^K\} \leqslant \lambda_j^{i,\min} + \overline{x}_j^i, \tag{5.65a}$$

$$x_j^{\max} = \max\{f_j(e) \mid e \in \boldsymbol{e}^K\} \geqslant \lambda_j^{i,\max} + \underline{x}_j^i, \tag{5.65b}$$

hence $\boldsymbol{\xi} \subseteq \boldsymbol{x}^H$, which was to be proved. □

Corollary 5.27 (Skalna and Hladík [259]) *Let* $\boldsymbol{x}^* := [x^*(e)]$ *and* $\boldsymbol{\xi}$ *be, respectively, an OI and IEH solution to problem (5.50). Then,*

$$\underline{x}_i^H \in [\underline{x}_j^*, \underline{\xi}_j], \quad \overline{x}_j^H \in [\overline{x}_j^*, \overline{\xi}_j]. \tag{5.66}$$

The p-solution obtained using the Interval-affine Gauss–Seidel Iteration permits us to determine an outer interval solution $\boldsymbol{x}^{OI}$, an inner estimation of hull solution $\boldsymbol{x}^{IEH}$, as well as intervals containing the endpoints of each component of hull solution $\boldsymbol{x}^{IH}$ related to system (4.10).

The sketch of the algorithm of the interval-affine Gauss–Seidel iteration (5.52) is presented in Algorithm 9. The stopping criterion of the iteration is based on formula (5.17). So, iteration (5.52) stops if distance $q\left([\boldsymbol{x}^i(e)], [\boldsymbol{x}^{i+1}(e)]\right) < \epsilon$, where ϵ is a tolerance parameter. Typically, $10^{-6} < \epsilon < 10^{-3}$, but this really depends on the problem being solved and the cost of the iterations. The asymptotic time complexity of the algorithm is $\mathcal{O}(n^3K + n^2mM)$, where n is the size of the problem, K is the number of uncertain parameters (noise symbols), m is the cost of the multiplication of revised affine forms, and M is the number of iterations taken by the algorithm.

Algorithm 9 (Interval-affine Gauss–Seidel Iteration)

Input: $A(\boldsymbol{p})x = b(\boldsymbol{p})$, $\boldsymbol{p}$, $\boldsymbol{x}$ (verified initial enclosure of $S(\boldsymbol{p})$), ϵ
Output: $\boldsymbol{x}(e)^*$ (p-solution)
  Transform $A(\boldsymbol{p})$, $b(\boldsymbol{p})$ into $\boldsymbol{C}(e)$, $\boldsymbol{c}(e)$, respectively
  $C^c = \text{mid}(\boldsymbol{C}(e))$; $R \approx (C^c)^{-1}$
  $\boldsymbol{V}(e) = R\boldsymbol{C}(e)$; $\boldsymbol{v}(e) = R\boldsymbol{c}(e)$
  if $\left(\rho\left(\left|(\boldsymbol{D}(e) + \boldsymbol{L}(e))^{-1}\,\boldsymbol{U}(e)\right|\right) < 1\right)$ **then**
    Transform $\boldsymbol{x}$ into $\boldsymbol{x}^0(e)$; $i = -1$
    repeat
      $i = i + 1$
      for $j = 1$ to n **do**
        $\boldsymbol{x}_j^{i+1}(e) = \frac{\boldsymbol{v}_j(e) - \sum_{\mu<j} \boldsymbol{V}_{j\mu}(e)\boldsymbol{x}_\mu^{i+1}(e) - \sum_{\mu>j} \boldsymbol{V}_{j\mu}(e)\boldsymbol{x}_\mu^{i}(e)}{\boldsymbol{V}_{jj}(e)}$
      end for
    until $\left(q\left([\boldsymbol{x}^i(e)], [\boldsymbol{x}^{i+1}(e)]\right) < \epsilon\right)$
    $\boldsymbol{x}^*(e) = \boldsymbol{x}^{i+1}(e)$
    return "Success"
  else
    return "Failure"
  end if

5.2.3 Convergence of Iterative Methods

In this section, we discuss the convergence properties for iterative methods that use revised affine forms. We consider a general class of interval operators, including Krawczyk [120], Gauss–Seidel, or Jacobi iterations as particular examples. Denote

$$\boldsymbol{W}(e) = \sum_{k=1}^{K} W^{(k)} e_k + \boldsymbol{W},$$

$$\boldsymbol{w}(e) = \sum_{k=1}^{K} w^{(k)} e_k + \boldsymbol{w},$$

$$\boldsymbol{x}(e) = \sum_{k=1}^{K} x^{(k)} e_k + \boldsymbol{x}.$$

Consider iterations

$$\boldsymbol{x}(e) \mapsto \boldsymbol{w}(e) + \boldsymbol{W}(e)\boldsymbol{x}(e), \tag{5.67}$$

where the right-hand side is computed by revised affine arithmetic. Consider also the standard multiplication of revised affine forms

$$\begin{aligned} \boldsymbol{W}(e)\boldsymbol{x}(e) = & \sum_{k=1}^{K} \left(W^{(k)} x^c + W^c x^{(k)}\right) e_k + W^c x^c + \\ & + \left(|W^c| x^{\Delta} + V^{\Delta} |x^c|\right) [-1, 1] + \\ & + \left(\sum_{k=1}^{K} |W^{(k)}| + W^{\Delta}\right) \left(\sum_{k=1}^{K} |x^{(k)}| + x^{\Delta}\right) [-1, 1]. \end{aligned}$$

Notice that term

$$\left(\sum_{k=1}^{K} |W^{(k)}| + W^{\Delta}\right) \left(\sum_{k=1}^{K} |x^{(k)}| + x^{\Delta}\right) [-1, 1]$$

is a simple enclosure for residual value

$$\boldsymbol{r}(e) = \left(\sum_{k=1}^{K} W^{(k)} e_k + W^{\Delta}[-1, 1]\right) \left(\sum_{k=1}^{K} x^{(k)} e_k + x^{\Delta}[-1, 1]\right),$$

and it can be replaced by tighter enclosures. However, the theory derived below holds for the simple enclosure as well as for any better one.

Denote

$$R = \sum_{k=1}^{K} |W^{(k)}| + W^{\Delta}.$$

Theorem 5.28 (Skalna and Hladík [259]) *If $\rho(R + |\boldsymbol{W}|) < 1$, then iterations (5.67) converge to a unique fixed point for each $\boldsymbol{x}(e)$.*

Proof Denote

$$\tilde{W}_x = \begin{pmatrix} W^c & & & W^{(1)} & \\ & \ddots & & \vdots & \\ & & W^c & V^{(K)} & \\ & & & W^c & \\ RD_{\operatorname{sgn}(x^{(1)})} & \dots & RD_{\operatorname{sgn}(x^{(K)})} & W^{\Delta}D_{\operatorname{sgn}(x^c)} & R + |W^c| \end{pmatrix},$$

$$\tilde{x} = \begin{pmatrix} x^{(1)} \\ \vdots \\ x^{(K)} \\ x^c \\ x^{\Delta} \end{pmatrix}, \quad \tilde{w} = \begin{pmatrix} w^{(1)} \\ \vdots \\ w^{(K)} \\ w^c \\ w^{\Delta} \end{pmatrix}, \quad \tilde{W} = \begin{pmatrix} |W^c| & & & |W^{(1)}| & \\ & \ddots & & \vdots & \\ & & |W^c| & |W^{(K)}| & \\ & & & |W^c| & \\ R & \dots & R & W^{\Delta} & R + |W^c| \end{pmatrix}.$$

The affine multiplication can be formulated as

$$\begin{aligned} \boldsymbol{V}(e)\boldsymbol{x}(e) = \sum_{k=1}^{K}(W^{(k)}x^c + W^c x^{(k)})e_k + W^c x^c + [-|W^c|, |W^c|x^{\Delta} + \\ + [-W^{\Delta}, W^{\Delta}]|x^c| + \sum_{k=1}^{K}[-R, R]x^{(k)} + [-R, R]x^{\Delta}. \end{aligned}$$

Thus, the iteration (5.67) can be reformulated as follows in terms of the coefficients of the revised affine forms

$$\tilde{x} \mapsto \tilde{w} + \tilde{W}_x\tilde{x} \subseteq \tilde{w} + [-\tilde{W}, \tilde{W}]\tilde{x}. \tag{5.68}$$

Now, due to the assumption $\rho(R + |\boldsymbol{W}|) < 1$ there is Perron vector $u > 0$, such that $(R + |\boldsymbol{W}|)^T u < 1$. Since

$$\begin{pmatrix} |W^c| & & & |W^{(1)}| & \\ & \ddots & & \vdots & \\ & & |W^c| & |W^{(K)}| & \\ & & & |W^c| & \\ R & \dots & R & W^{\Delta} & R + W^{\Delta} \end{pmatrix}^T \begin{pmatrix} u \\ \vdots \\ u \end{pmatrix} \leq \begin{pmatrix} (R + |\boldsymbol{W}|)^T u \\ \vdots \\ (R + |\boldsymbol{W}|)^T u \end{pmatrix} < \begin{pmatrix} u \\ \vdots \\ u \end{pmatrix},$$

the spectral radius of $\tilde{W}$ is strictly less than 1.

Now, consider two affine forms $\boldsymbol{x}(e)$ and $\boldsymbol{y}(e)$, which are equivalently represented by their coefficient vectors $\tilde{x}$ and $\tilde{y}$, respectively. For their images, it holds

$$|\tilde{w} + \tilde{W}_x\tilde{x} - \tilde{w} - \tilde{W}_y\tilde{y}| = |\tilde{W}_x\tilde{x} - \tilde{W}_y\tilde{y}|.$$

Due to the property

$$RD_{\mathrm{sgn}(x)}x - RD_{\mathrm{sgn}(y)}y = R|x| - R|y| = R(|x| - |y|) \leq R|x - y|$$

we have

$$|\tilde{W}_x\tilde{x} - \tilde{W}_y\tilde{y}| \leq \tilde{W}|\tilde{x} - \tilde{y}|.$$

Therefore, by [5], the iterations form the so called $\tilde{W}$-contraction, and converge for each initial setting. □

The above iterations involved the Krawczyk type of iterative methods. Now, we extend the results to also cover Jacobi and Gauss–Seidel iterations. Let

$$\boldsymbol{U}(e) = \sum_{k=1}^{K} U^{(k)}e_k + \boldsymbol{U},$$

and consider iterations

$$\boldsymbol{x}(e) \mapsto \boldsymbol{U}(e)\left(\boldsymbol{w}(e) + \boldsymbol{W}(e)\boldsymbol{x}(e)\right). \tag{5.69}$$

For any matrix in affine form $\boldsymbol{W}(e)$, denote

$$R_W = \sum_{k=1}^{K} |W^{(k)}| + W^{\Delta}.$$

Theorem 5.29 (Skalna and Hladík [259]) *Iterations (5.69) converge if*

$$\rho(R_U + |\boldsymbol{U}|) < 1 \quad \textit{and} \quad \rho(R_W + |\boldsymbol{W}|) < 1.$$

Proof It follows from Theorem 5.28 and its proof by applying it twice on the contracting interval operations

$$\boldsymbol{x}(e) \mapsto \boldsymbol{y}(e) = \boldsymbol{w}(e) + \boldsymbol{W}(e)\boldsymbol{x}(e), \quad \boldsymbol{y}(e) \mapsto \boldsymbol{x}(e) = \boldsymbol{U}(e)\boldsymbol{y}(e).$$

□

Consider the alternative iterations to (5.69)

$$\boldsymbol{x}(e) \mapsto \boldsymbol{U}(e)\boldsymbol{w}(e) + (\boldsymbol{U}(e)\boldsymbol{W}(e))\boldsymbol{x}(e), \tag{5.70}$$

and further denote $\boldsymbol{G}(e) = \boldsymbol{U}(e)\boldsymbol{W}(e)$. Now, it directly follows that

Corollary 5.30 *Iterations (5.70) converge if $\rho(R_G + |\boldsymbol{G}|) < 1$.*

For system (5.50), which reads $\boldsymbol{V}(e)x = \boldsymbol{v}(e)$, the Krawczyk iteration [120] has the particular form of

$$\boldsymbol{x}(e) \mapsto \boldsymbol{v}(e) + (I_n - \boldsymbol{V}(e))\boldsymbol{x}(e).$$

We can generalize the properties of the Krawczyk iteration as follows:

Theorem 5.31 (Skalna and Hladík [259]) *Let $\boldsymbol{x}(e)$ be such that*

$$\boldsymbol{v}(e) + (I_n - \boldsymbol{V}(e))\boldsymbol{x}(e) \subseteq \operatorname{int} \boldsymbol{x}(e) \tag{5.71}$$

for each $e \in \boldsymbol{e}^n$. Then,

(i) $\boldsymbol{V}(e)$ is regular for each $e \in \boldsymbol{e}^n$,
(ii) $\boldsymbol{x}(e)$ is an enclosure for the solution set of (5.50) for each $e \in \boldsymbol{e}^n$.

Proof It directly follows from the properties of the Krawczyk iteration for standard interval systems of equations applied to a fixed $e \in \boldsymbol{e}^n$; see [234]. □

In order to employ this observation, we need to check for (5.71). The following lines give a recipe for how to do this efficiently:

Proposition 5.32 (Skalna and Hladík [259]) *Let $\boldsymbol{x}(e)$, $\boldsymbol{y}(e)$ be given. Then, $\boldsymbol{x}(e) \subseteq \operatorname{int} \boldsymbol{y}(e)$ for each $e \in \boldsymbol{e}^n$ if and only if*

$$\overline{x} + \sum_{k=1}^{K} |x^{(k)} - y^{(k)}| < \overline{y},$$
$$\underline{x} - \sum_{k=1}^{K} |x^{(k)} - y^{(k)}| > \underline{y}.$$

Proof The maximum value of

$$\boldsymbol{x}(e) - \boldsymbol{y}(e) = \sum_{k=1}^{K} (x^{(k)} - y^{(k)})e_k + \boldsymbol{x} - \boldsymbol{y}$$

is attained for $x = \overline{x}$, $y = \underline{y}$ and $e_k = \operatorname{sgn}(x^{(k)} - y^{(k)})$. Its value must be negative, from which the first condition follows. The second condition is proved analogously. □

5.2.4 Generalized Expansion Method

Degrauwe et al. [36] proposed a method for solving linear systems of form (4.25). The method is called the *Expansion Method* (EM), since it is based on the expansion of the inverse of a matrix in the Neumann series. The formulae for the first and second order Expansion Method can be found in [36]. Below, we present a modified version of the Expansion Method, which employs arithmetic on revised affine forms and involves the residual correction of the right-hand vector. The proposed modifications have several advantages. The use of the arithmetic of revised affine forms enables us to obtain an arbitrary-order expansion and produce a p-solution, the advantages of which have already been described. The improved version of the Expansion method will be referred to as the *Generalized Expansion Method* (GEM).

Let us consider system (4.25). It is not hard to see that vector x solves this system if and only if $u = x - \tilde{x}$ solves system

$$\boldsymbol{C}(e)u = \boldsymbol{c}(e) - \boldsymbol{C}(e)\tilde{x}. \tag{5.72}$$

The pre-conditioning of system (5.72) with $\left(C^{(0)}\right)^{-1}$ yields new system

$$\boldsymbol{V}(e)u = \boldsymbol{v}(e), \tag{5.73}$$

where

$$\begin{aligned}
\boldsymbol{V}(e) &= I + \sum_{k=1}^{K} V^{(k)}\varepsilon_k + \boldsymbol{W},\\
V^{(k)} &= \left(C^{(0)}\right)^{-1} C^{(k)} \ (k = 1, \ldots, K),\\
\boldsymbol{W} &= \left|\left(C^{(0)}\right)^{-1}\right| C^{\Delta}[-1, 1],
\end{aligned}$$

$$\begin{aligned}
\boldsymbol{v}(e) &= v^{(0)} + \sum_{k=1}^{K} v^{(k)}\varepsilon_k + \boldsymbol{w},\\
v^{(k)} &= \left(C^{(0)}\right)^{-1}\left(b^{(k)} - C^{(k)}\tilde{x}\right), \ (k = 0, \ldots, K),\\
\boldsymbol{w} &= \left|\left(C^{(0)}\right)^{-1}\right|\left(c^{\Delta} + C^{\Delta}|\tilde{x}|\right)[-1, 1].
\end{aligned}$$

If $V(e) \in \boldsymbol{V}(e)$, then there is $e \in \boldsymbol{e}$ and $W \in \boldsymbol{W}$ such that

$$V(e) = I + \sum_{k=1}^{K} V^{(k)}\varepsilon_k + W.$$

If $\rho\left(\sum_{k=1}^{K}\left|V^{(k)}\right| + W^{\Delta}\right) < 1$, then the inverse of $V(e)$ can be expressed by the Neumann series, i.e.,

$$V(e)^{-1} = I + \sum_{i=1}^{\infty}(-1)^i\left(\sum_{k=1}^{K}V^{(k)}\varepsilon_k + W\right)^i.$$

It is not hard to see (cf. [36]) that, for $i \geqslant 1$,

$$(-1)^i\left(\sum_{k=1}^{K}V^{(k)}\varepsilon_k + W\right)^i \in \left(V^{\Delta} + W^{\Delta}\right)^i[-1, 1],$$

where $V^{\Delta} = \sum_{k=1}^{K}\left|V^{(k)}\right|$. So, we have

$$\begin{aligned}V(e)^{-1} &\in I + \sum_{i=1}^{m}(-1)^i\left(\sum_{k=1}^{K}V^{(k)}\varepsilon_k + W\right)^i + \\ &+ \left(V^{\Delta} + W^{\Delta}\right)^{m+1}\left(I + \sum_{i=1}^{\infty}\left(V^{\Delta} + W^{\Delta}\right)^i\right)[-1, 1] = \\ &= I + \sum_{i=1}^{m}(-1)^i\left(\sum_{k=1}^{K}V^{(k)}\varepsilon_k + W\right)^i + \\ &+ \left(V^{\Delta} + W^{\Delta}\right)^{m+1}\left(I - \left(V^{\Delta} + W^{\Delta}\right)\right)^{-1}[-1, 1],\end{aligned}$$

which means that

$$\begin{aligned}\boldsymbol{V}(e)^{-1} \subseteq I &+ \sum_{i=1}^{m}(-1)^i\left(\sum_{k=1}^{K}V^{(k)}\varepsilon_k + \boldsymbol{W}\right)^i + \\ &+ \left(V^{\Delta} + W^{\Delta}\right)^{m+1}\left(I - \left(V^{\Delta} + W^{\Delta}\right)\right)^{-1}[-1, 1].\end{aligned}$$

Put

$$\boldsymbol{g}(e) \triangleq \boldsymbol{H}(e)\boldsymbol{v}(e),$$

where

$$\begin{aligned}\boldsymbol{H}(e) = I &+ \sum_{i=1}^{m}(-1)^i\left(\sum_{k=1}^{K}V^{(k)}\varepsilon_k + \boldsymbol{W}\right)^i + \\ &+ \left(V^{\Delta} + W^{\Delta}\right)^{m+1}\left(I - \left(V^{\Delta} + W^{\Delta}\right)\right)^{-1}[-1, 1].\end{aligned} \tag{5.74}$$

Then, the p-solution of system (4.25) is given by

$$x(e) = Le + x, \tag{5.75}$$

where $L = \left(g^{(1)} \dots g^{(K)}\right) \in \mathbb{R}^{n \times K}$ and $x = \tilde{x} + g^{(0)} + g$.

Proposition 5.33 *Let $x(e)$ be defined by (5.75). Then, the solution set of system (4.25) is included in $[x(e)]$.*

Proof If x is a solution to system (4.25), then there is $C(e) \in C(e)$ and $c(e) \in c(e)$ such that $C(e)x = c(e)$. Hence, $u = x - \tilde{x}$ is a solution to system (5.72) and, thus, to system (5.73) as well. So,

$$x \in \tilde{x} + V(e)^{-1}v(e) \subseteq \tilde{x} + H(e)v(e) \subseteq [x(e)].$$

□

The inner estimate of the hull solution can be obtain using the formulae analogous to formulae (5.58) and (5.59) (see Sect. 5.2.2).

Obviously, higher-order expansions should give more-accurate results, but they are also more-time-consuming. By using a simple example, we show the performance of the Generalized Expansion Method. The results of the original approach from [36] are included as well. In order to simplify the notation, in what follows the order of the expansion methods will be given in parentheses. For example, GEM(2) will denote GEM method of second order.

Example 5.34 Consider the parametric interval linear system with affine linear dependencies given by

$$\begin{pmatrix} p_1 + p_6 & -p_6 & 0.0 & 0.0 & 0.0 \\ -p_6 & p_2 + p_6 + p_7 & -p_7 & 0.0 & 0.0 \\ 0.0 & -p_7 & p_3 + p_7 + p_8 & -p_8 & 0.0 \\ 0.0 & 0.0 & -p_8 & p_4 + p_8 + p_9 & -p_9 \\ 0.0 & 0.0 & 0.0 & -p_9 & p_5 + p_9 \end{pmatrix} \begin{pmatrix} v_1 \\ v_2 \\ v_3 \\ v_4 \\ v_5 \end{pmatrix} = \begin{pmatrix} 10 \\ 0 \\ 10 \\ 0 \\ 0 \end{pmatrix},$$

with $p_k \in [1 - \delta, 1 + \delta]$, $k = 1, \dots, 9$. The system is obtained while solving the Okumura's problem of a linear resistive network (see, e.g., [168, 173]). By using the affine transformation, we obtain the system of the form required by the Expansion Method. The results for $\delta = 0.1$ obtained using the Expansion and Generalized Expansion methods are presented in Table 5.4 (cf. Fig. 5.3). It can be seen that the first-order GEM method is better than the second-order EM method. The third-order expansion slightly improves the bounds, but the further increase of expansion m order has a negligible effect.

Table 5.4 Comparison of result obtained using EM and GEM methods for Example 5.34 ($\delta = 0.1$)

EM(1)	EM(2)	GEM(1)	GEM(2)	GME(5)	GME(10)
[6.028, 8.153]	[6.171, 8.062]	[6.298, 8.013]	[6.300, 8.006]	[6.302, 8.003]	[6.302, 8.003]
[3.230, 5.134]	[3.372, 5.023]	[3.484, 4.953]	[3.487, 4.948]	[3.489, 4.945]	[3.489, 4.945]
[4.548, 6.361]	[4.672, 6.268]	[4.806, 6.215]	[4.809, 6.208]	[4.811, 6.206]	[4.811, 6.206]
[1.499, 2.865]	[1.592, 2.783]	[1.688, 2.717]	[1.693, 2.712]	[1.694, 2.710]	[1.694, 2.710]
[0.586, 1.596]	[0.657, 1.531]	[0.726, 1.472]	[0.731, 1.467]	[0.733, 1.466]	[0.733, 1.466]

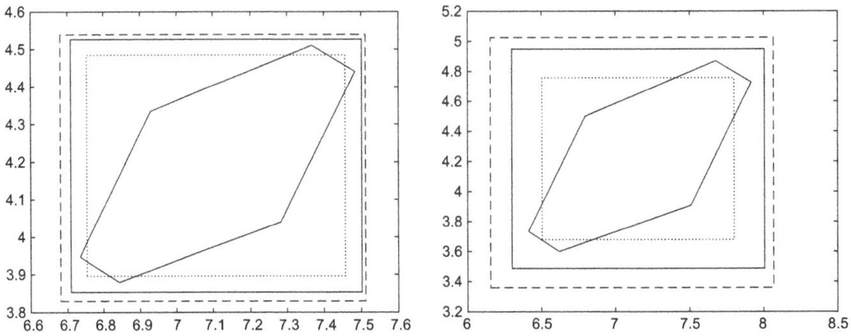

Fig. 5.3 Projection of parametric solution set from Example 5.34 on v_1v_2-plane ($\delta = 0.05$ [left], $\delta = 0.1$ [right]), and OI solution obtained using Expansion method (dashed line), OI solution obtained using GEM method (solid line), and IEH solution obtained using GEM method (dotted line)

5.3 Methods for Interval Hull Solution

For parametric interval linear systems with affine-linear dependencies such that uncertainties are present only in the right-hand-side vector, the interval hull solution can be obtained as stated in the following proposition:

Proposition 5.35 *Let $Ax = b(\boldsymbol{p})$, $\boldsymbol{p} \in \mathbb{IR}^K$ be a parametric system such that $A \in \mathbb{R}^{n\times n}$ is non-singular, and let x^c be a solution to the mid-point system. Then*

$$\text{hull}\,(S(\boldsymbol{p})) = x^c + \sum_{k=1}^{K} A^{-1} b^{(k)} \boldsymbol{p}_k.$$

In the general case, we must find the tightest box that encloses the parametric solution set (cf. (5.2)). As already discussed, if $A(p) \in A(\boldsymbol{p})$ is non-singular, then $x_i = x_(p)$ $(i = 1, \ldots, n)$. We recall that is assumed throughout that x_i satisfy certain continuity conditions. Obviously, the endpoints of the tightest interval enclosing Rge $(x_i \mid \boldsymbol{p})$ are equal to the extremal values of $x_i(p)$ over $\boldsymbol{p}$. Hence, the problem of computing the interval hull solution of a parametric interval linear system (4.10) can be formulated as the problem of solving the following $2n$ constrained optimization problems: for $i = 1, \ldots, n$,

$$\text{minimize } x_i(p) \qquad \text{st. } A(p)x(p) = b(p),\ p \in \boldsymbol{p}, \tag{5.76a}$$

$$\text{maximize } x_i(p) \qquad \text{st. } A(p)x(p) = b(p),\ p \in \boldsymbol{p}. \tag{5.76b}$$

Proposition 5.36 *Let $\underline{x}_i$ and $\overline{x}_i$ denote the solution to the ith minimization (5.76a) and maximization (5.76b) problem, respectively. Let $\boldsymbol{x} = ([\underline{x}_1, \overline{x}_i] \ldots [\underline{x}_n, \overline{x}_n])^T$. Then,*

$$\text{hull}\,(S(\boldsymbol{p})) \equiv \boldsymbol{x}.$$

However, the explicit formula for $x_i(p)$ is very difficult to obtain in the general case. Therefore, we need a method that will be able to compute the extrema of a real-valued function that is given implicitly by a system of parametric linear equations.

5.3.1 Parametric Interval Global Optimization

In (Skalna [252]), we proposed a method that can be used to compute the interval hull solution of parametric interval linear systems with both affine-linear and non-linear dependencies. This method, called *Parametric interval global optimization* (PIGO), combines the Moore–Skeelboe algorithm for global optimization [146, 261], with a method for solving parametric interval linear systems. By using the latter method to bound the range of $x_i(p)$ $(i = 1, \ldots, n)$ over $\boldsymbol{p}$, we obtain a *fail-safe procedure* [74], i.e., we are guaranteed that the boxes that contain a global optimum are never deleted.

The basic scheme of the interval global optimization method (see Algorithm 10) is quite simple. We start from the initial box $\boldsymbol{p}$ in which the global minimum is sought. We split the initial box into sub-boxes and discard those that cannot contain the global minimum. All feasible boxes are stored in sorted list L (in increasing order with respect to the value of a function to be optimized) for further processing. At nth iteration list L consists of n pairs

$$L = \{(v_{n1}, \boldsymbol{v}_{n1}), \ldots, (v_{nn}, \boldsymbol{v}_{nn})\},$$

where v_{nj} equals to the minimal value of an objective function on $\boldsymbol{v}_{nj}$ $(j = 1, \ldots, n)$. The leading pair in list L is denoted by $(y_n, \boldsymbol{y}_n)$ and boxes $\boldsymbol{y}_n$ are called the *leading boxes* of the optimization algorithm.

Theorem 5.37 (Ratschek and Rokne [197]) *Let $f : D \to \mathbb{R}$, $D \subseteq \mathbb{R}^n$, $\boldsymbol{x} \subseteq D$, and let $\boldsymbol{f}$ be an inclusion function for f. Then, global minimum $f^* = \inf\,(\text{Rge}\,(f \mid \boldsymbol{x}))$ exists and satisfies*

$$f^* \in \boldsymbol{f}(\boldsymbol{y}_n)$$

Algorithm 10 (Parametric interval global optimization)

Input: $A(\boldsymbol{p})x = b(\boldsymbol{p})$, $\boldsymbol{p}$, ϵ
Output: $\boldsymbol{x}_i^{IH}$ computed with some arbitrary precision $\epsilon > 0$
  Compute outer solution $\boldsymbol{x}^{OI}$ to system $A(\boldsymbol{p})x = b(\boldsymbol{p})$
  Put $y = \inf\left(\boldsymbol{x}_i^{OI}\right)$, $\boldsymbol{y} = \boldsymbol{p}$
  Initialize list $L = \{(y, \boldsymbol{y})\}$ and cut-off level $z = \sup\left(\boldsymbol{x}_i^{OI}\right)$
  repeat
    Remove first element $(y, \boldsymbol{y})$ from list L
    if $(\mathrm{wid}(\boldsymbol{y}) < \epsilon)$ **then**
      return $(y, \boldsymbol{y})$
    end if
    Choose subdivision direction ν
    Bisect $\boldsymbol{y}$ along direction ν such that $\boldsymbol{y} = \boldsymbol{v}^1 \cup \boldsymbol{v}^2$
    Compute outer solutions $\boldsymbol{x}^{OI,1}$ and $\boldsymbol{x}^{OI,2}$ to systems
    $A(\boldsymbol{v}^1)x = b(\boldsymbol{v}^1)$ and $A(\boldsymbol{v}^2)x = b(\boldsymbol{v}^2)$, respectively
    Put $v^1 = \inf\left(\boldsymbol{x}_i^{OI,1}\right)$, $v^2 = \inf\left(\boldsymbol{x}_i^{OI,2}\right)$,
    $z = \min\left\{z, \sup\left(\boldsymbol{x}_i^{OI,1}\right), \sup\left(\boldsymbol{x}_i^{OI,2}\right)\right\}$
    Discard box $\boldsymbol{v}^j$ with $v^j > z$ $(j = 1, 2)$ // (cut-off test)
    Perform the monotonicity test
    For each feasible box, insert pair $(v^j, \boldsymbol{v}^j)$ accordingly (so that the list remains increasingly sorted by the function value) into list L
    Denote the first on the list by $(y, \boldsymbol{y})$
  until $(L = \{\})$

for all leading boxes $\boldsymbol{y}_n$ *of the optimization algorithm.*

In order to effectively discard infeasible sub-boxes, we use the following acceleration techniques (cf. [32, 74, 197, 252]).

Range check or "midpoint test". Let f be a function to be optimized, $\boldsymbol{x}$ a given box, and let $\overline{f}$ denote the current best rigorous upper bound on the global minimum. Compute the lower bound $\underline{f}$ on $\mathrm{Rge}\,(f \mid \boldsymbol{x})$. If $\underline{f} > \overline{f}$, then $\boldsymbol{x}$ cannot contain the global minimum; thus, it can be discarded.

Rules for selecting subdivision direction. Csendes and Ratz discussed four subdivision rules in [32] that use the following *merit function*:

$$k = \min\left\{j \mid j \in \{1, \ldots, K\} \text{ and } D(j) = \max_{i=1}^{n} D(i)\right\}, \tag{5.77}$$

where $D(i)$ is determined by a given rule. In order to avoid the necessity of computing the derivatives, we used only two of the four rules in [252].

Rule A. The first rule, called the *interval-width-oriented rule* [146, 197, 261], chooses the coordinate direction with

$$D(i) = x_i^{\Delta}. \tag{5.78}$$

The uniform subdivision causes the width of the actual sub-boxes to go the quickest to zero [32].

Rule B. The second rule aims at decreasing the excess width between the computed bounds and Rge $(f \mid \boldsymbol{x})$ caused, in part, by rounding errors [32]. The rule is defined by (5.77) with

$$D(i) = \begin{cases} x_i^{\Delta}, & 0 \in \boldsymbol{x}_i, \\ x_i^{\Delta}/\mathrm{mig}(\boldsymbol{x}_i), & \text{otherwise.} \end{cases} \tag{5.79}$$

Monotonicity test. The monotonicity test (MT) is used to check whether a function to be optimized is strictly monotone in box $\boldsymbol{y}$. If so, the box cannot contain the minimizer in its interior and, hence, can be discarded or reduced to one of its endpoints. For the problem of computing the interval hull solution, the monotonicity check can be performed using one of the methods described in Sect. 5.1.6.

Algorithm 10 presents the modified version of the Moore–Skelboe algorithm for computing ith component of the hull of a parametric solution set.

Lemma 5.38 (Ratschek and Rokne [197]) *Let* $(y_n, \boldsymbol{y}_n)_{n=1}^{\infty}$ *be a sequence of pairs generated by the interval global optimization algorithm with subdivision rule A. Then,* $\mathrm{wid}(\boldsymbol{y}_n) \to 0$ *as* $n \to \infty$.

So, in order to guarantee the convergence of Algorithm 10 it is enough to choose inclusion function $\boldsymbol{f}$ of function f to be optimized that satisfies condition

$$\mathrm{wid}(\boldsymbol{f}(\boldsymbol{y})) \to 0 \text{ as } \mathrm{wid}(\boldsymbol{y}) \to 0. \tag{5.80}$$

Theorem 5.39 (Ratschek and Rokne [197]) *If inclusion function* $\boldsymbol{f}$ *satisfies condition (5.80), then sequence* $(\boldsymbol{f}(\boldsymbol{y}_n))_{n=1}^{\infty}$ *converges to global minimum* f^**, i.e., sequence* $(y_n)_{n=1}^{\infty}$ *converges to* f^* *from below.*

For the problem of computing the hull solution, the inclusion function for x_i is given by outer solution $\boldsymbol{x}^{OI}$, which has the required property. Namely, if $\boldsymbol{p} \to 0$, then system $A(\boldsymbol{p})x = b(\boldsymbol{p})$ tends to the point system; thus, $\mathrm{wid}(\boldsymbol{x}^{OI}) \to 0$.

Example 5.40 Consider the following parametric interval linear system:

$$\begin{pmatrix} p_1 & p_2+1 & -2p_1 \\ p_2+1 & -3 & p_1+p_2 \\ 2-p_1 & 4p_2+1 & 1 \end{pmatrix} \begin{pmatrix} x_1 \\ x_2 \\ x_3 \end{pmatrix} = \begin{pmatrix} 2p_1 \\ p_2-1 \\ -1 \end{pmatrix},$$

where $p_1 \in [-2.0-\delta, -1.0+\delta]$, $p_2 \in [3.0-\delta, 4.0+\delta]$, and $\delta = 0.5, 0.6$. The results obtained using different variants of the global optimization algorithm are presented in Table 5.5, whereas Fig. 5.4 shows the boxes processed while computing the upper bound for $\boldsymbol{x}_1^{OI}$.

As we can see in this case, the performance of the PIGO does not depend on the subdivision selection rule, but does depend a great deal on the monotonicity test.

Table 5.5 Comparison of performance of different variants of PIGO for Example 5.40

PIGO	#Box	CPU [s]
Rule A−MT	447	0.4
Rule B−MT	481	0.4
Rule A+MT	35	0.2
Rule B+MT	35	0.2

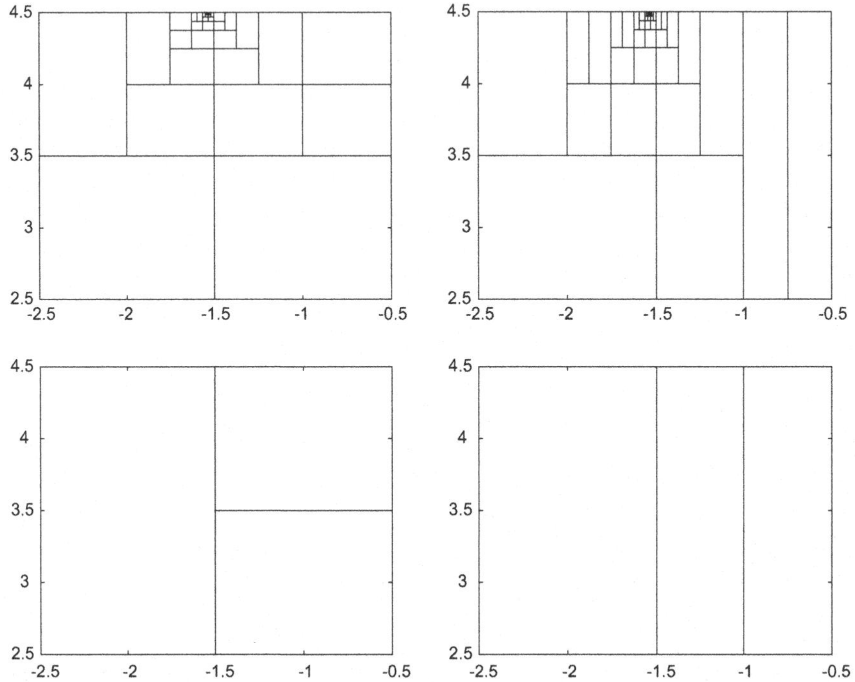

Fig. 5.4 Boxes processed while computing upper bound for $\boldsymbol{x}_1^{OI}$ in Example 5.40. Boxes processed by PIGO without monotonicity test (Rule A [top left], Rule B [top right]) and with monotonicity test (Rule A [bottom left], Rule B [bottom right])

Thanks to the use of the latter, the number of processed boxes has been reduced ten times, and the overall computational time has been halved.

5.3.2 *Evolutionary Strategy*

In a series of papers [41–43, 249, 256, 257], we proposed several metaheuristic strategies to obtain approximate solutions for problems (5.76a) and (5.76b), and subsequently, to obtain an inner estimate of the hull solution. The main advantage of the metaheuristic-based approaches over the methods described in the previous sections is that the matrix of the system does not have to be strongly regular. This means that the metaheuristic-based approaches enable us to solve a larger class of problems than any of the remaining methods. Another advantage of the metaheuristic-based approaches is that they can be used in a straightforward manner to solve problems with arbitrary dependencies.

The *evolution strategy* (ES) was the first metaheuristic optimization that we used to compute the hull solution of a parametric interval linear system [249]. Further research has shown that ES is one of the best metaheuristic strategies for the considered problem. In our evolution strategy, population P consists of N individuals characterized by vectors

$$p^i = \left(p^i_1, \ldots, p^i_K\right)^T, \; i = 1, \ldots, N,$$

where $p^i_k \in \boldsymbol{p}_k, k = 1, \ldots K$. The initial population is generated at random based on a uniform distribution. Then, the population evolves over T generations according to the following strategy. A number (N_x) of the best individuals from a given generation pass to the next generation, and the rest of the new population is obtained by using

- *tournament selection*: a tournament is run on two or more individuals, and the "strongest" one is chosen for the process of reproduction. In order to decide which individual is "stronger," we must solve a system of linear equations, therefore, the overall method is generally quite costly. However, it is still worth considering because of its undoubted advantages,
- *arithmetic crossover*: parents p^i, p^j selected in a tournament are recombined, with probability π_c, which results in two offspring

$$\begin{aligned} \left(p^i\right)' &= rp^i + (1-r)p^j, \\ \left(p^j\right)' &= rp^j + (1-r)p^i. \end{aligned}$$

  Here, r is a random number from interval $[0, 1]$,
- *non-uniform mutation*: each offspring $\left(p^i\right)', \left(p^j\right)'$ mutates (with a probability π_m) according to formula

$$\left(p^i\right)'_k = \begin{cases} \left(p^i\right)'_k + \left(\overline{p}^i_k - \left(p^i\right)'\right)\left(1 - r^{(1-\frac{t}{T})b}\right), & \text{if } q < 0.5 \\ \left(p^i\right)'_k + \left(\left(p^i\right)'_k - \underline{p}^i_k\right)\left(1 - r^{(1-\frac{t}{T})b}\right), & \text{if } q \geqslant 0.5, \end{cases},$$

  where t is the number of the current generation, $k \in \{1, \ldots, K\}$ is a random index, $r, q \in [0, 1]$ are random numbers, and b is a system parameter determining the degree of dependency on the generation number. The non-uniform mutation keeps the population from stagnating in the early stages of the evolution and tunes the solution in the later stages of evolution.

The values of the ES parameters depend on the problem to be solved. However, based on a great many experiments, we can conclude that mutation rate π_m should be greater than 0.5 and the value of b should be greater than 2, whereas crossover rate π_c should be less than 0.3. Population of size ≈80 and the number of generations ≈60 guarantee good inner estimates of the hull solution. As already discussed, inner estimates are very useful in assessing the quality of outer enclosures. The general outline of the our evolutionary strategy is presented in Algorithm 11.

Algorithm 11 (Evolutionary strategy for computing the lower bound of an inner estimate of the hull solution)

Input: $A(\boldsymbol{p})x = b(\boldsymbol{p})$, $\boldsymbol{p}$,
  ES parameters: T, N, N_x, b, π_c, π_m
Output: $\underline{x}_i \geqslant \inf\left(\boldsymbol{x}_i^{IH}\right)$
  Initialize $P = \{p^1, \ldots, p^N\}$, where $p^i \in \boldsymbol{p}$ are vectors of random values
  $t = 0$
  while $(t < T)$ **do**
    Select N_x best individuals from P that pass to the next generation
    $N_t = N_x$
    while $(N_t \leqslant N)$ **do**
      Select parents p^i and p^j from P using tournament selection
      if $(r_{[0,1]} < \pi_c)$ **then**
        Set $\left(p^i\right)' = p^i$, $\left(p^j\right)' = p^j$
        Recombine p^i and p^j → offspring $\left(p^i\right)'$, $\left(p^j\right)'$
      end if
      if $(r_{[0,1]} < r_m)$ **then**
        Mutate $\left(p^i\right)'$
      end if
      if $(r_{[0,1]} < r_m)$ **then**
        Mutate $\left(p^j\right)'$
      end if
      $N_t = N_t + 2$
    end while
  end while

Example 5.41 Consider the parametric interval linear system from Example 5.40. The parameters of the ES are set as follows: $N = 10$, $T = 10$, $b = 6$, $N_x = 0.1N$, $\pi_c = 0.2$, $\pi_m = 0.7$, and the tournament size is 4. The obtained results are resented in Table 5.6. As we can see, the ES produced a very good inner estimate of the hull solution.

As we can see, the results of ES and PIGO coincide up to fourth decimal places. In this case, the computational effort of both methods is comparable, however for larger problems the PIGO is inefficient, whereas the ES is still able to produce good inner estimations of the hull solution in a reasonable amount of time.

Table 5.6 Result of evolutionary strategy for Example 5.40: average result (ES ave.) from 100 runs; standard deviations for each endpoint (std.dev. lb, std.dev.ub); and result of parametric interval global optimization

PIGO	ES ave.	std.dev. lb	std.dev. ub
[0.303, 0.724]	[0.303, 0.724]	0	0.001
[−0.197, −0.128]	[−0.196, −0.128]	0.003	0
[−0.746, −0.011]	[−0.746, −0.011]	0	0

5.4 Comparison of Enclosure Methods

In this section we compare the enclosure methods in terms of speed and accuracy. Speed is an important factor in solving large systems of equations, because for larger problems the volume of computations involved is huge. Whereas accuracy is of great importance for better knowledge on the behavior of a system. The presented here examples include among others some practical problems of structural and electrical engineering. All computations were performed using our own software, which implements operations on intervals and on revised affine forms in `binary64` floating-point arithmetic. The operating system is MS Windows 10 Pro and the compiler is VC++12.0.

The quality of outer interval enclosures will be assessed by using the following measures. Given two intervals $\boldsymbol{x}$, $\boldsymbol{y}$ such that $\boldsymbol{x} \subseteq \boldsymbol{y}$

- *standard overestimation measure*

$$O_w(\boldsymbol{x}, \boldsymbol{y}) = \left(1 - \frac{x^{\Delta}}{y^{\Delta}}\right) 100\%, \tag{5.81}$$

- *sharpness measure* [189]

$$O_s(\boldsymbol{x}, \boldsymbol{y}) = \begin{cases} 1, & y^{\Delta} = 0, \\ 0, & \boldsymbol{x} = \emptyset, \\ \frac{x^{\Delta}}{y^{\Delta}}, & \text{otherwise.} \end{cases} \tag{5.82}$$

If the hull solution is available, then it can be used to assess the quality of outer enclosures. However, as discussed earlier, the problem of computing the hull solution is NP-hard. That is why, outer enclosure should be compared with inner estimate of the hull solution, which is provided with a little additional effort by methods that are based strictly on revised affine forms. The obtained results are compared with the results from the literature, whenever the latter are available.

5.4.1 Parametric Systems with Affine-Linear Dependencies

Example 5.42 In [185] it is argued that the Direct Method (which is called in [185] the Bauer–Skeel method) yields guaranteed enclosures only in exact-precision rational arithmetic. As an example of a failure of the Direct Method, the following parametric interval linear system was given in [185]:

$$\begin{pmatrix} p_1 & p_1 \\ p_1 & p_1 + 0.01 \end{pmatrix} \cdot \begin{pmatrix} x_1 \\ x_2 \end{pmatrix} = \begin{pmatrix} p_2 \\ p_2 + 0.01 \end{pmatrix},$$

where $p_1 \in [0.9, 1.1]$, $p_2 \in [1.9, 2.1]$ $(\rho(B^{\Delta} \approx 0.1,\ \operatorname{cond}(A^c) \approx 402.01)$. We present below (Table 5.7) the results obtained using our implementation of selected methods described in previous sections. For the comparison purposes, we present as well the results reported in [185]. They were obtained after one initial iteration of Krawczyk iteration (KI) and 15 refinement iterations (with threshold $\delta = 10^{-16}$). As we can see, all the results are guaranteed, since they enclose the exact bounds, which are $\left(\left[\frac{8}{11}, \frac{4}{3}\right], [1, 1]\right)$. The PHBRC method turned out to be the best in this case.

Table 5.7 Comparison of enclosures for Example 5.42

IAGSI
[0.6969601796260643,1.333333333333577]
[0.9999999999997983,1.000000000000145]
MRFPI
[0.6639999999998135,1.336000000000243]
[0.9999999999998003,1.000000000000143]
PIGSI
[0.6969601796260645,1.333333333333576]
[0.9999999999998003,1.000000000000144]
DM
[0.6666666666664817,1.333333333333575]
[0.9999999999998003,1.000000000000143]
PHBRC
[0.7272727272726000,1.333333333333575]
[0.9999999999998004,1.000000000000143]
GEM(2)
[0.6969370252518848,1.333333333333575]
[0.9999999999998003,1.000000000000143]
KI
[0.6666666666661077, 1.33333333338497]
[0.9999999999995374, 1.00000000004625]

Table 5.8 Comparison of outer interval enclosures for Example 5.43: outer interval solution $\boldsymbol{x}^{OI}$ and average overestimation $O_w(\boldsymbol{x}^{OI}, \boldsymbol{x}^H)$ of hull solution $\boldsymbol{x}^H$

Method	$\boldsymbol{x}_1^{OI}$	$O_w(\boldsymbol{x}_1^{OI}, \boldsymbol{x}_1^H)$ (%)	$\boldsymbol{x}_2^{OI}$	$O_w(\boldsymbol{x}_2^{OI}, \boldsymbol{x}_2^H)$ (%)
MRFPI, DM	[−3, 12]	41	[−24, 18]	46
PHBRC	[2, 12]	12	[−24, −0.667]	3
NP	[−5.87, 14.87]	57	[−34.23, 28.23]	64
GEM(3)	[−0.12980, 12]	27	[−24, 9.1129]	31
PGSI	[−1.27266, 12]	33	[−24, 8.0088]	29
IAGSI	[−0.2035, 12]	28	[−24, 6.45]	25
PIGO (33 #Iter)	[3.1579, 12]	0	[−24, −1.2632]	0

Example 5.43 Consider following parametric interval linear system (cf. [163]):

$$\begin{pmatrix} p_1 & \frac{1}{2} - p_2 \\ 1 + p_1 & p_2 \end{pmatrix} \begin{pmatrix} x_1 \\ x_2 \end{pmatrix} = \begin{pmatrix} 6 \\ 6 \end{pmatrix}, \tag{5.83}$$

where $p_1, p_2 \in [0.5, 1.5]$ ($\rho(B^{\Delta}) \approx 0.82$, $\mathrm{cond}(A^c) \approx 2.76$). In [182] were given the results of Neumaier–Pownuk (NP) method [163], the DM method, and the MRFPI. These results are presented Table 5.8 together with the result obtained using the methods from the previous sections. The average overestimation of the hull is included as well. Using a simple computation it can be shown that the extremal values of the solutions $x_1(p_1, p_2)$, $x_2(p_1, p_2)$ are attained at the vertices of the box $[0.5, 1.5] \times [0.5, 1.5]$. The combinatorial approach yields the following exact bounds (see Fig. 5.5):

$$\boldsymbol{x}^H = \begin{pmatrix} [3.15789, 12] \\ [-24, -1.263157] \end{pmatrix}.$$

Table 5.8 shows that the PIGO produced optimal result after 33 iterations. From among other methods the PHBRC method turned out to be the best and the IAGSI method took the second place. In this case PBS method produced the same result as the DM and MRFPI methods. All the methods, except the NP method, produced exact lower bound for x_2 and exact upper bound for x_1.

Example 5.44 Consider a parametric interval linear system with tridiagonal matrix (cf. [189])

$$A(p) = \begin{pmatrix} 1 & p_1 & & & \\ p_1 & 1 & p_1 & & \\ & p_1 & \ddots & \ddots & \\ & & \ddots & \ddots & p_1 \\ & & & p_1 & 1 \end{pmatrix}, \quad b(p) = (-p_1, 0, \ldots, 0, -p_2)^T, \tag{5.84}$$

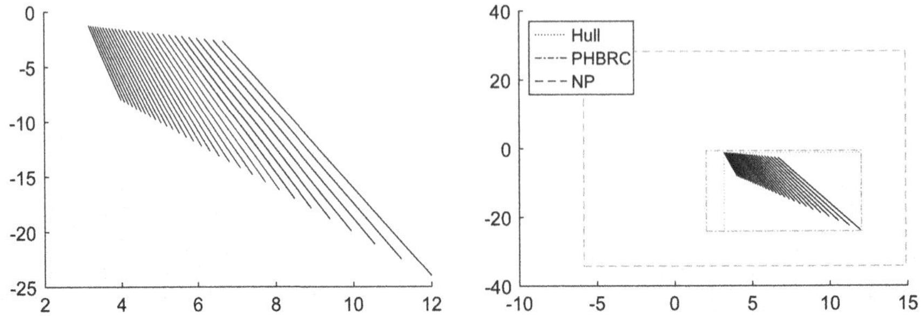

Fig. 5.5 Solution set of system from Example 5.43 (left); solution set with hull solution and NP and PHBRC enclosures (right)

Table 5.9 Sharpness of outer enclosures for Example 5.44; minimal and maximal values of overestimation measured based on inner estimate of hull solution

Method		IAGSI	MRFPI	PIGSI	DM, PHBRC	GEM(2)
n	δ	Min–max	Min–max	Min–max	Min–max	Min–max
5	0.1	0.93–1.00	0.89–1.00	0.90–1.00	0.90–1.00	0.93–1.00
5	1	0.47–0.99	0.36–0.98	0.37–0.99	0.37–0.99	0.47–0.99
5	10	0.00–0.87	0.00–0.80	0.00–0.84	0.00–0.81	0.00–0.87
50	0.1	0.99–1.00	0.99–1.00	0.99–1.00	0.99–1.00	0.99–1.00
50	1	0.98–0.99	0.97–0.99	0.98–0.99	0.97–0.99	0.98–0.99
50	10	0.81–0.90	0.78–0.85	0.81–0.89	0.78–0.85	0.81–0.89
100	0.1	0.99–1.00	0.99–1.00	0.99–1.00	0.99–1.00	0.99–1.00
100	1	0.98–0.99	0.97–0.99	0.97–0.99	0.97–0.99	0.98–0.99
100	10	0.82–0.88	0.74–0.82	0.79–0.88	0.76–0.84	0.81–0.88
200	0.1	0.99–1.00	0.99–1.00	0.99–1.00	0.99–1.00	0.99–1.00
200	1	0.97–0.98	0.95–0.97	0.96–0.98	0.95–0.97	0.97–0.98
200	10	0.70–0.77	0.60–0.67	0.65–0.72	0.61–0.69	0.69–0.76

where $p_1 \in [100 - \delta, 100 + \delta]$, $p_2 \in [1 - \delta/100, 1 + \delta/100]$. We solve the system for different dimensions and different tolerances. Similarly as in [189], we use the sharpness measure to assess the quality of the obtained interval enclosures. The inner estimation of the hull, produced by the IAGSI method, is used for the comparison purposes. The results are presented in Table 5.9.

Table 5.9 shows that the IAGSI and GEM(2) produced the sharpest enclosures, and the advantage over the remaining methods increases along with the amount of uncertainty.

Table 5.10 reports the computational times and the number of iterations taken by the iterative methods. As we can see, the DM method achieved the best performance.

Table 5.10 CPU times (in seconds) for Example 5.44; number of iterations taken by iterative methods is given in parentheses

n	δ	IAGSI	MRFPI	PIGSI	DM	PHBRC	GEM(2)
5	0.1	0.009(3)	0.005(1)	0.010(3)	0.004	0.003	0.01
5	1	0.011(4)	0.005(2)	0.010(4)	0.004	0.003	0.01
5	10	0.017(6)	0.005(2)	0.010(6)	0.004	0.003	0.01
50	0.1	2.88(3)	3.05(1)	2.76(2)	0.69	2.08	2.56
50	1	3.31(3)	3.11(1)	2.78(2)	0.69	2.08	2.56
50	10	3.75(4)	3.14(1)	2.91(3)	0.69	2.08	2.56
100	0.1	19.62(3)	29.19(1)	19.26(2)	3.62	15.47	17.56
100	1	20.31(4)	29.36(1)	19.28(2)	3.62	15.47	17.56
100	10	20.61(5)	31.12(2)	19.45(3)	3.62	15.47	17.56
200	0.1	144.84(4)	227.41(1)	141.16(2)	23.94	120.43	126.52
200	1	160.36(5)	242.97(1)	145.79(2)	23.94	120.43	126.52
200	10	167.06(7)	248.042(2)	153.59(4)	23.94	120.43	126.52

Example 5.45 Consider a parametric interval linear system with affine-linear dependencies defined by [37]:

$$\begin{aligned} A^{(k)} &= (k+1)\cdot L, \quad k = 0, \ldots, K, \\ b^{(k)} &= 1, \end{aligned} \tag{5.85}$$

where $L \in \mathbb{R}^n$ is the `Lehmer` matrix defined, for $i, j = 1, \ldots, n$, by

$$L(i, j) = \begin{cases} i/j, \ i \leqslant j, \\ j/i, \ i > j. \end{cases}$$

The parameters of the system are subjected to tolerances $p_k \in [1-\delta, 1+\delta]$, $k = 1, \ldots, K$. The system is solved for various n, K and δ. The values of sharpness measure and the computational times are presented in Tables 5.11 and 5.12, respectively. As we can see from the tables, the GEM(2) method generally shows the best performance.

Example 5.46 Consider the following over-determined parametric linear system (cf. [177, 281])

$$\begin{pmatrix} p_1 & p_1+1 & p_1+2 & p_1 \\ p_1 & p_1+2 & p_1+3 & p_1+1 \\ p_1+1 & p_1+2 & p_1+3 & p_1+2 \\ p_1+2 & p_1+3 & p_1+4 & p_1+3 \\ p_1+3 & p_1+4 & p_1+5 & p_1+5 \\ p_1+5 & p_1+5 & p_1+6 & p_1+7 \end{pmatrix} \begin{pmatrix} x_1 \\ x_2 \\ x_3 \\ x_4 \end{pmatrix} = \begin{pmatrix} p_2 \\ 1 \\ -2 \\ -3 \\ 0 \\ 0 \end{pmatrix},$$

Table 5.11 Sharpness of outer enclosures for Example 5.45

n	K	δ	IAGSI	MRFPI	PIGSI	DM, PHBRC	GEM(2)
10	10	0.3	0.71–0.72	0.59–0.60	0.71–0.72	0.61–0.62	0.72–0.73
10	20	0.05	0.95–0.96	0.92–0.93	0.95–0.96	0.92–0.93	0.95–0.96
10	20	0.3	0.70–0.71	0.58–0.59	0.70–0.71	0.60–0.61	0.71–0.72
50	10	0.3	0.71–0.72	0.59–0.60	0.71–0.72	0.61–0.62	0.72–0.73
50	20	0.05	0.95–0.96	0.92–0.93	0.95–0.96	0.92–0.93	0.95–0.96
50	20	0.3	0.70–0.71	0.58–0.59	0.70–0.71	0.60–0.61	0.71–0.72
100	10	0.3	0.71–0.72	0.59–0.60	0.71–0.72	0.61–0.62	0.72–0.73
100	20	0.05	0.95–0.96	0.92–0.93	0.95–0.96	0.92–0.93	0.95–0.96
100	20	0.3	0.70–0.71	0.58–0.59	0.70–0.71	0.60–0.61	0.71–0.72
100	20	0.4	0.59–0.60	0.47–0.48	0.59–0.60	0.48–0.49	0.59–0.60
100	20	0.5	0.47–0.48	0.34–0.35	0.47–0.48	0.37–0.38	0.46–0.47

Table 5.12 Comparison of computational times and number of iterations (in parentheses) for Example 5.45

n	K	δ	IAGSI	MRFPI	PIGSI	DM	PHBRC	GEM(2)
10	10	0.3	0.128(2)	0.0711(2)	0.126(2)	0.115	0.047	0.11
10	20	0.05	0.199(2)	0.109(1)	0.194(2)	0.145	0.067	0.12
10	20	0.3	0.212(2)	0.117(2)	0.201(2)	0.145	0.067	0.12
50	10	0.3	23.39(2)	9.41(2)	19.79(2)	11.19	6.14	6.25
50	20	0.05	24.95(2)	15.27(1)	22.98(2)	12.19	7.14	7.25
50	20	0.3	25.05(2)	15.72(2)	23.55(2)	12.19	7.14	7.25
100	10	0.3	115.76(2)	103.84(2)	128.43(2)	87.74	42.91	40.37
100	20	0.05	181.92(2)	113.24(1)	184.94(2)	118.43	63.24	69.35
100	20	0.3	182.87(2)	124.47(2)	182.92(2)	118.43	63.24	69.35
100	20	0.4	193.94(2)	127.15(2)	192.15(2)	133.27	63.24	69.35
100	20	0.5	195.91(2)	131.61(3)	194.73(2)	133.27	63.24	69.35

The nominal value of parameter $p_1 = 1$ ($\rho(B^{\Delta}) = 0.25$, condition number of the respective matrix (4.39) is $\approx$1246). The matrix of the system is a full rank matrix ($\text{rank}(A(p)) = 4$) and the right-hand side vector $b(p)$ is constructed so that for every p_1 and $p_2 = 3$ the exact solution of system $A(p)^T A(p)x = A(p)^T b$ is $x = (1, -1, 1, -1)^T$ (see [282]). We solve the parametric interval linear system with $p_1 \in [0.9, 1.1]$ and $p_2 \in [2.995, 3.005]$. The obtained enclosures are presented in Table 5.13 (the results are given with 16-digits mantissa).

As we can see, all the considered methods produced guaranteed sharp enclosures for the parametric solution set.

Table 5.13 Comparison of enclosures for Example 5.46

IAGSI	PIGSI
[0.9...97938, 1.0...0214]	[0.9...98235, 1.0...0178]
[−1.0...01677, −0.9...983227]	[−1.0...01383, −0.9...986561]
[0.9...989102, 1.0...01119]	[0.9...906386, 1.0...09521]
[−1.0...0397, −0.9...9608]	[−1.0...03522, −0.9...964517]
MRFPI	DM
[0.9...98180, 1.0...0186]	[0.9...98235, 1.0...0177]
[−1.0...01400, −0.9...985773]	[−1.0...01383, −0.9...986561]
[0.9...90092, 1.0...0977]	[0.9...90638, 1.0...0952]
[−1.0...0370, −0.9...96376]	[−1.0...0352, −0.9...96451]
PHBRC	GEM(2)
[0.9...98266, 1.0...0174]	[0.9...98332, 1.0...0171]
[−1.0...01338, −0.9...987008]	[−1.0...01249, −0.9...98730]
[0.9...90907, 1.0...0925]	[0.9...91258, 1.0000...00860]
[−1.0...0345, −0.9...96521]	[−1.0...0342, −0.9...9666]

Table 5.14 Sharpness of enclosures for Example 5.47

IAGSI	MRFPI	PIGSI	DM, PHBRC	GEM(2)
0.53–0.73	0.41–0.65	0.46–0.72	0.43–0.66	0.52–0.73

5.4.2 Parametric Systems with Nonlinear Dependencies

One of the main advantages of the proposed in this book approach is that it can be used straightforwardly to solve PILS with nonlinear dependencies.

Example 5.47 Consider parametric interval linear system

$$\begin{pmatrix} -(p_1+p_2)p_3 & p_1+3 \\ p_3 & p_1p_2p_3 \end{pmatrix} \begin{pmatrix} x_1 \\ x_2 \end{pmatrix} = \begin{pmatrix} p_3^2 \\ 0 \end{pmatrix},$$

where $p_1, p_2 \in [0.9, 1.1]$, $p_3 \in [1.9, 2.1]$ ($\rho(B^{\Delta}) \approx 0.21$, $\text{cond}(A^c) = 2$).

Table 5.14 presents the comparison of the obtained interval enclosures. As we can see, the IAGSI and GEM(2) methods turned out to be the best, with a small advantage of the former.

The solution set with the inner and outer bounds produced by the IAGSI method (see Table 5.15) is shown in Fig. 5.6.

Table 5.15 Inner and outer bounds of IAGSI method for Example 5.47

Outer	Inner
[−0.568693, −0.418154]	[−0.533088, −0.453759]
[0.367412,0.643581]	[0.404181,0.606812]

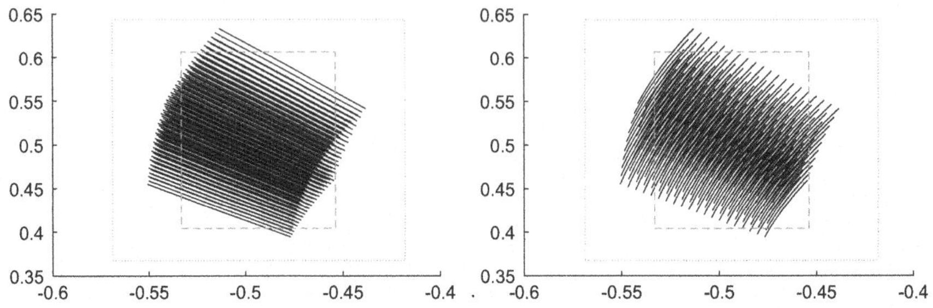

Fig. 5.6 Solution set of system from Example 5.47 depicted as set of plane curves depending on parameter p_3 (left) and as set of plane curves depending on parameter p_2 (right); rectangles represent inner and outer bounds obtained using IAGSI method

Table 5.16 Sharpness of enclosures for Example 5.48

	IAGSI	MRFPI	PIGSI	DM, PHBRC	GEM(2)
Min–max	0.69–0.91	0.61–0.89	0.67–0.89	0.63–0.88	0.69–0.91

Example 5.48 Consider parametric interval linear system (cf. [46])

$$\begin{pmatrix} -(p_1+p_2)p_2 & p_1p_3 & p_2 \\ p_2p_4 & p_2^2 & 1 \\ p_1p_2 & p_3p_5 & \sqrt{p_2} \end{pmatrix} \begin{pmatrix} x_1 \\ x_2 \\ x_3 \end{pmatrix} = \begin{pmatrix} 1 \\ 1 \\ 1 \end{pmatrix},$$

where $p_1 \in [1, 1.2]$, $p_2 \in [2, 2.2]$, $p_3 \in [0.5, 0.51]$, $p_4, p_5 \in [0.39, 0.4]$ ($\rho(B^{\Delta}) = 0.14$, $\text{cond}(A^c) \approx 4.02$). Table 5.16 presents the sharpness of enclosures produced by the considered methods. As we can see, the IAGSI method and the GEM(2) method produced the best enclosures. Table 5.17 compares the interval bounds produced by the IAGSI method and the bounds reported in [46].

Example 5.49 Consider parametric interval linear system

$$\begin{pmatrix} p_1^2(p_2+p_3)-2 & p_1p_2^2 & p_1^3 & -4 & p_3^3 \\ -p_1^2p_2+4 & p_1^2+p_2^2 & 3+p_3^2 & 3p_3p_4-1 & 0 \\ (p_1-p_3)p_2 & 3 & p_2p_3 & p_1p_2p_5 & 1+p_5 \\ p_4p_5-p_1 & p_2^3+p_4 & p_2p_3 & p_2p_3p_4+p_5^2 & -p_4 \\ p_4^2+1 & -p_4 & p_3^2 & p_2p_3p_4+p_5^2 & -p_2^2 \end{pmatrix} \begin{pmatrix} x_1 \\ x_2 \\ x_3 \\ x_4 \\ x_5 \end{pmatrix} = \begin{pmatrix} p_1 \\ p_1^2-p_2p_3 \\ -2p_3 \\ -2 \\ 1 \end{pmatrix},$$

Table 5.17 Comparison of interval bounds of IAGSI method and bounds reported in [46] for Example 5.48

IAGSI				[46]	
Outer		Inner		Outer	
0.044077209	0.049580275	0.044935774	0.048721709	0.043718642	0.049772302
0.074711517	0.087283627	0.075980387	0.086014757	0.074017025	0.087572793
0.582922989	0.626895908	0.584912863	0.624906035	0.581819347	0.627210871

Table 5.18 Sharpness of outer enclosures for Example 5.49

δ	IAGSI	MRFPI	PIGSI	DM, PHBRC	GEM(2)
	Min–max	Min–max	Min–max	Min–max	Min–max
0.01	0.76–0.88	0.68–0.87	0.72–0.84	0.72–0.84	0.76–0.88
0.03	0.31–0.59	0.23–0.55	0.25–0.51	0.25–0.51	0.31–0.59
0.05	0.00–0.19	0.00–0.15	0.00–0.14	0.00–0.14	0.00–0.16

Table 5.19 Comparison of computational times and number of iterations (in parentheses) for Example 5.49

δ	IAGSI	MRFPI	PIGSI	DM	PHBRC	GEM(2)
0.01	0.021(6)	0.012(2)	0.014(5)	0.011	0.011	0.015
0.03	0.035(10)	0.013(2)	0.023(9)	0.011	0.011	0.015
0.05	0.049(19)	0.016(4)	0.036(18)	0.011	0.011	0.015

where $p_1 \in 4\boldsymbol{\delta}$, $p_2 \in 1.8\boldsymbol{\delta}$ $p_3 \in 2.5\boldsymbol{\delta}$, $p_4 \in 2.5\boldsymbol{\delta}$, $p_5 \in 1.0\boldsymbol{\delta}$, $\boldsymbol{\delta}= [1-\delta, 1+\delta]$. The results for $\delta = 0.01, 0.03, 0.05$ ($\rho(B^{\Delta}) \approx 0.13, 0.4, 0.68$, $\operatorname{cond}(A^c) = 64.53$) are presented in Tables 5.18, 5.19 and 5.20.

Example 5.50 Consider parametric interval linear system

$$\begin{pmatrix} 1/p_1 - 2 & \sqrt{p_1 p_2^2} & p_1^3 & -4 \\ -p_1^2 p_2 + 4 & 1/(p_1^2 + p_2^2) & \sqrt{3 + p_3} & 3p_3 p_4 - 1 \\ 3 & (p_1 - p_3)p_2 & \sqrt{p_2 p_3} & p_1 p_2 p_5 \\ p_4 p_5 - p_1 & (2p_4 - p_3)^2 p_2^2 & p_2 p_3 & p_2 p_3 p_4 + p_5^2 \end{pmatrix} \begin{pmatrix} x_1 \\ x_2 \\ x_3 \\ x_4 \end{pmatrix} = \begin{pmatrix} p_1 \\ p_1^2 - p_2 p_3 \\ -2p_3 \\ -2 \end{pmatrix},$$

where $p_1 \in 1.2\boldsymbol{\delta}$, $p_2 \in 0.8\boldsymbol{\delta}$, $p_3 \in 0.51\boldsymbol{\delta}$, $p_4 \in 2.51\boldsymbol{\delta}$, $p_5 \in 1.01\boldsymbol{\delta}$, $\boldsymbol{\delta}= [1-\delta, 1+\delta]$ ($\rho(B^{\Delta}) \approx 0.03, 0.06, 0.29, 0.61$, $\operatorname{cond}(A^c) \approx 10.8$). The results for $\delta = 0.005$, $0.01, 0.05, 0.1$ are presented in Tables 5.21, 5.22 and 5.23.

As we can see, also in this case, the IAGSI and GEM methods produced the sharpest enclosures.

Table 5.20 Inner and outer bounds from IAGSI method for Example 5.49

$\delta = 0.01$		$\delta = 0.03$	
[−0.938575, −0.844690]	[−0.927497, −0.855768]	[−1.095471, −0.708888]	[−0.961619, −0.842740]
[−0.761835, −0.596390]	[−0.748259, −0.609965]	[−0.994711, −0.369152]	[−0.830499, −0.533364]
[1.326755, 1.501491]	[1.343484, 1.484762]	[1.091169, 1.783339]	[1.296936, 1.577572]
[−0.668166, −0.527430]	[−0.659164, −0.536432]	[−0.866545, −0.355143]	[−0.757656, −0.464032]
[−1.461549, −1.160048]	[−1.443130, −1.178467]	[−1.872297, −0.787433]	[−1.648709, −1.011021]

Table 5.21 Sharpness of outer enclosures for Example 5.50

δ	IAGSI	MRFPI	PIGSI	DM	PHBRC	GEM
	Min–max	Min–max	Min–max	Min–max	Min–max	Min–max
0.005	0.93–0.96	0.92–0.95	0.93–0.96	0.92–0.95	0.92–0.95	0.93–0.96
0.01	0.88–0.92	0.86–0.90	0.86–0.92	0.85–0.90	0.85–0.90	0.87–0.92
0.05	0.41–0.57	0.36–0.49	0.37–0.55	0.36–0.50	0.36–0.55	0.40–0.56
0.1	0.00–0.07	0.00–0.05	0.00–0.06	0.00–0.05	0.00–0.07	0.00–0.06

Table 5.22 Comparison of computational times and number of iterations (in parentheses) for Example 5.50

δ	IAGSI	MRFPI	PIGSI	DM	PHBRC	GEM
0.005	0.012(4)	0.005(1)	0.013(3)	0.004	0.005	0.009
0.01	0.012(4)	0.009(1)	0.015(4)	0.004	0.005	0.009
0.05	0.021(7)	0.009(2)	0.017(6)	0.004	0.005	0.009
0.1	0.024(12)	0.009(3)	0.024(11)	0.004	0.005	0.009

Summarizing the results from this section and the previous one, it can be seen that generally the IAGSI and GEM methods are superior to the other methods in terms of the quality of produced enclosures. The PIGSI method yields slightly worse enclosures, next are DM and PHBRC methods, and finally the MRFPI method. However, in some cases the PHBRC might be superior to other methods. So, it seems to be a good idea to solve a system using one the methods which "always" yield good enclosures and intersect the obtained result with the result of the PHBRC method. When it comes to the CPU time, the DM, PHBRC and GEM(2) methods were generally the most efficient.

5.4.3 Practical Applications

The case studies presented in this section illustrate practical applications of the methods described in the previous sections. We consider small but realistic problems coming from structural and electrical engineering.

Example 5.51 Consider a simple one-bay structural steel frame, shown in Fig. 5.7, which was initially analyzed by Corliss et al. [30]. By applying conventional methods for frame structures analysis, the following parametric linear system is obtained (cf. [30, 178])

Table 5.23 Inner and outer bounds from IAGSI method for Example 5.50

$\delta = 0.01$		$\delta = 0.05$	
[−0.65464, −0.63212]	[−0.65344, −0.63331]	[−0.71648, −0.57129]	[−0.67753, −0.61024]
[−0.18451, −0.16042]	[−0.1835, −0.16143]	[−0.25416, −0.10405]	[−0.22152, −0.13669]
[1.06453, 1.1044]	[1.06692, 1.10201]	[0.95381, 1.22076]	[1.03193, 1.14264]
[0.30676, 0.33794]	[0.30821, 0.3365]	[0.22457, 0.42046]	[0.27109, 0.37393]

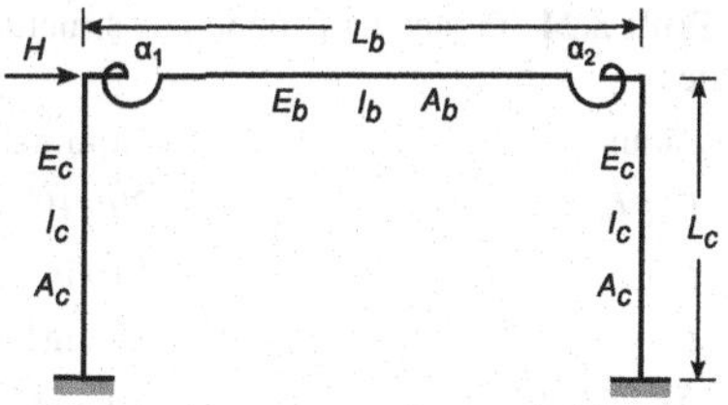

Fig. 5.7 One-bay structural
steel frame (cf. [30, 178])

$$
\left(
\begin{array}{ccccc}
\frac{A_b E_b}{L_b} + \frac{12 E_c I_c}{L_c^3} & 0 & \frac{6 E_c I_c}{L_c^2} & 0 & 0 \\[4pt]
0 & \frac{A_c E_c}{L_c} + \frac{12 E_b I_b}{L_b^3} & 0 & \frac{6 E_b I_b}{L_b^2} & \frac{6 E_b I_b}{L_b^2} \\[4pt]
\frac{6 E_c I_c}{L_c^2} & 0 & \alpha + \frac{4 E_c I_c}{L_c} & -\alpha & 0 \\[4pt]
0 & \frac{6 E_b I_b}{L_b^2} & -\alpha & \alpha + \frac{4 E_b I_b}{L_b} & \frac{2 E_b I_b}{L_b} \\[4pt]
0 & \frac{6 E_b I_b}{L_b^2} & 0 & \frac{2 E_b I_b}{L_b} & \alpha + \frac{4 E_c I_c}{L_c} \\[4pt]
-\frac{A_b E_b}{L_b} & 0 & 0 & 0 & 0 \\[4pt]
0 & -\frac{12 E_b I_b}{L_b^3} & 0 & -\frac{6 E_b I_b}{L_b^2} & -\frac{6 E_b I_b}{L_b^2} \\[4pt]
0 & & & &
\end{array}
\right.
$$

(5.86)

$$
\left.
\begin{array}{ccc}
-\frac{A_b E_b}{L_b} & 0 & 0 \\[4pt]
0 & -\frac{12 E_b I_b}{L_b^3} & 0 \\[4pt]
0 & 0 & 0 \\[4pt]
0 & -\frac{6 E_b I_b}{L_b^2} & 0 \\[4pt]
0 & -\frac{6 E_b I_b}{L_b^2} & -\alpha \\[4pt]
\frac{A_b E_b}{L_b} + \frac{12 E_c I_c}{L_c^3} & 0 & \frac{6 E_c I_c}{L_c^2} \\[4pt]
0 & \frac{A_c E_c}{L_c} + \frac{12 E_b I_b}{L_b^3} & -\frac{6 E_b I_b}{L_b^2} \\[4pt]
\frac{6 E_c I_c}{L_c^2} & -\frac{6 E_b I_b}{L_b^2} & \alpha + \frac{4 E_c I_c}{L_c}
\end{array}
\right)
\begin{pmatrix} d2_x \\ d2_y \\ r2_z \\ r5_z \\ r6_z \\ d3_x \\ d3_y \\ r3_z \end{pmatrix}
=
\begin{pmatrix} H \\ 0 \\ 0 \\ 0 \\ 0 \\ 0 \\ 0 \\ 0 \end{pmatrix}.
$$

As can be seen, the elements of system (5.86) are rational functions of Young modulus E_b, E_c, second moment of area I_b, I_c, cross-sectional area A_b, A_c, length L_b, L_c and joint stiffness α. The right-hand vector depends only on horizontal force H. In [30] all the parameters, except the lengths, were assumed to be uncertain and varying within given intervals. The nominal values of the model parameters and the worst case uncertainties are given in Table 5.24.

In order to compare the results produced by different methods, we solve system (5.86) with parameter uncertainties which are 1% of the values presented in the last column of Table 5.24 (cf. [30, 178]). Similarly as in [178], we assess the quality of the obtained enclosures by using O_w measure (see Table 5.25). Additionally, we provide the results from [178], obtained by combining parametric fixed-point iteration method with generalized interval arithmetic.

As can be seen from Table 5.25, the IAGSI method turned out to be the best from among the considered methods. Moreover, for solution components $r2_z$, $r5_z$, $r6_z$, it produced significantly better enclosures than the method from [178]. The next table

Table 5.24 Parameters of one-bay structural steel frame: nominal values and worst case uncertainties

Parameter	Nominal value	Uncertainty
E_b, E_c	$29 \cdot 10^6$ lbs/in^2	$\pm 348 \cdot 10^4$
I_b	510 in^4	± 51
I_c	272 in^4	± 27.2
A_b	10.3 in^2	± 1.3
A_c	14.4 in^2	± 1.44
H	5305.5 lbs	± 2203.5
α	$2.77461 \cdot 10^8$ lb-in/rad	$\pm 1.26504 \cdot 10^8$
L_b	288 in	
L_c	144 in	

Table 5.25 Comparison of interval enclosures and computational times (last row) for one-bay steel frame

Solution component	IAGSI (%)	MRFPI (%)	PIGSI (%)	DM, PHBRC (%)	GEM(2) (%)	[178] (%)
$d2_x$	6.44	6.83	6.62	6.66	6.62	1.66
$d2_y$	4.80	5.08	4.92	4.99	4.92	1.15
$r2_z$	6.77	7.28	7.02	7.11	7.02	9.17
$r5_z$	7.74	8.50	8.21	8.31	8.21	17.31
$r6_z$	7.80	8.62	8.28	8.42	8.28	27.08
$d3_x$	6.52	6.79	6.66	6.70	6.66	1.68
$d3_y$	4.96	5.28	5.14	5.20	5.14	1.59
$r3_z$	6.79	7.21	7.04	7.16	7.04	6.80
Time [s]	0.045	0.021	0.037	0.015, 0.011	0.039	

(Table 5.26) shows the hull solution (which can be obtained by using combinatorial approach) and the results of the IAGSI method. All results are rounded to 9 decimal places.

Example 5.52 Consider the plane four-bay two-floor truss structure depicted in Fig. 5.8. The truss is subjected to equal downward forces of 20 kN at nodes 2, 3, and 4, and has fixed supports at nodes 1 and 5.

The conventional Finite Element Method (FEM) for trusses gives the following parametric system of equations

$$K(p)x = F, \tag{5.87}$$

where p is a vector of model parameters, $K(p)$ is a stiffness matrix, F is a vector of nodal loads, and u is a vector of nodal displacements. The variation in the loading is $\pm 10\%$ from the nominal value. Each beam has the cross-sectional area $A = 0.005\,\text{m}^2$;

Table 5.26 Hull solution and results of IAGSI method for one-bay steel frame example; results are rounded outwardly to 10 digits accuracy

Solution component	Hull		IAGSI			
			Outer		Inner	
$d2_x$	0.152233722	0.154306458	0.1521984365	0.1543401680	0.152267432	0.154271172
$d2_y \times 10^3$	0.323803830	0.329780617	0.3237217862	0.3298459651	0.323869186	0.329698564
$r2_z \times 10^3$	−0.971680260	−0.957697225	−0.9719161507	−0.9574434577	−0.971426474	−0.957933134
$r5_z \times 10^3$	−0.469077618	−0.462295657	−0.4692073694	−0.4621523551	−0.468934350	−0.462425374
$r6_z \times 10^3$	−0.430183324	−0.423871097	−0.4303081289	−0.4237397067	−0.430051984	−0.423995851
$d3_x$	0.149693607	0.151738954	0.1496583315	0.1517726221	0.149727273	0.151703680
$d3_y \times 10^3$	−0.677375565	−0.664489801	−0.6775223003	−0.6643087801	−0.677194472	−0.664636608
$r3_z \times 10^3$	−0.939613234	−0.925977009	−0.9398470717	−0.9257314869	−0.939367696	−0.926210862

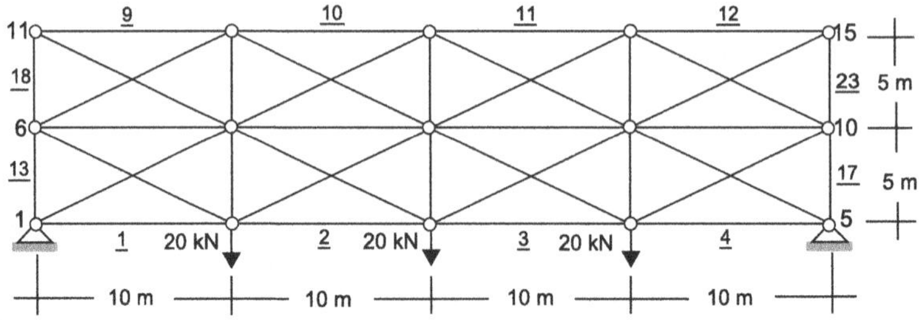

Fig. 5.8 Four-bay two-floor truss structure

and uncertain modulus of elasticity is varying within a tolerance interval around the nominal value $E = 2.0 \cdot 10^{11}$ Pa (the modulus of elasticity of each element are assumed to be varied independently). The relative uncertainty in the modulus of elasticity is varied to be able to assess the degradation of the bounds as the uncertainty increases. The lengths of the horizontal and vertical beams are, respectively, $L_H =$ 10 m and $L_V = 5$ m. There are 15 nodes, 38 elements and 3 loads, which results in 26 variables and 41 uncertain parameters. The width of the tolerance interval for the modulus of elasticity is varied to be able to assess the degradation of the bounds for displacements as the uncertainty increases. The sharpness measure is used to assess th quality of the outer interval enclosures. The results are presented in Tables 5.27 and 5.28.

In this case, the IAGSI and GEM(2) methods produced similar results, however GEM(2) turned out to be a bit faster.

Example 5.53 Consider parametric interval linear system (5.88), which occurs in worst-case tolerance analysis of linear DC (direct current) electrical circuits (cf. [40,

Table 5.27 Sharpness of outer enclosures for Example 5.52

Uncert. (%)	IAGSI	MRFPI	PIGSI	DM, PHBRC	GEM
	Min–max	Min–max	Min–max	Min–max	Min–max
1	0.43–0.92	0.43–0.92	0.43–0.92	0.43–0.92	0.43–0.92
2	0.25–0.84	0.24–0.83	0.24–0.83	0.24–0.83	0.25–0.84
3	0.11–0.75	0.10–0.74	0.10–0.74	0.10–0.74	0.11–0.75

Table 5.28 Comparison of computational times and number of iterations (in parentheses) for Example 5.52

Uncert. (%)	IAGSI	MRFPI	PIGSI	DM	PHBRC	GEM(2)
1	1.28(5)	0.939(2)	0.72(3)	0.38	0.39	0.928
2	1.41(6)	0.939(2)	0.74(4)	0.38	0.39	0.928
3	1.48(7)	0.939(2)	0.84(5)	0.38	0.39	0.928

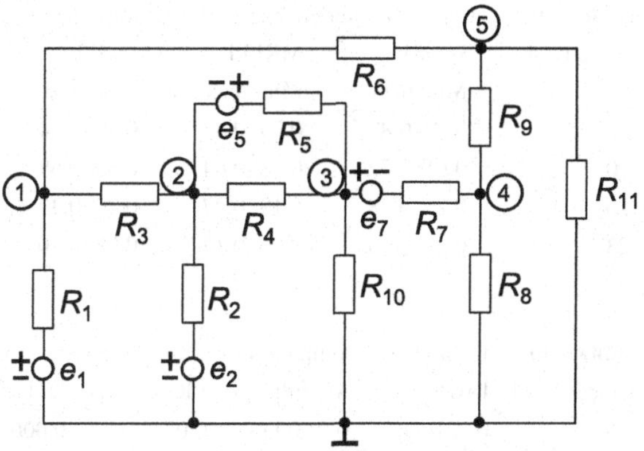

Fig. 5.9 Linear electrical circuit with five nodes and eleven branches

102, 105, 283]). The circuit studied is shown in Fig. 5.9 (cf. [102]). It has eleven branches and five nodes. The goal here is to find bounds for the node voltages $V_1, \ldots, V_5$. The nominal values of the resistances are $R_i = 100\,\Omega$, $i = 1, \ldots, 11$, the source voltages are $e_1 = e_2 = 100\,\text{V}$, $e_5 = e_7 = 10\,\text{V}$.

$$\begin{pmatrix} \frac{1}{R_1}+\frac{1}{R_3}+\frac{1}{R_6} & -\frac{1}{R_3} & 0 & 0 & -\frac{1}{R_6} \\ -\frac{1}{R_3} & \frac{1}{R_2}+\frac{1}{R_3}+\frac{1}{R_4}+\frac{1}{R_5} & -\frac{1}{R_4}-\frac{1}{R_5} & 0 & 0 \\ 0 & -\frac{1}{R_4}-\frac{1}{R_5} & \frac{1}{R_4}+\frac{1}{R_5}+\frac{1}{R_7}+\frac{1}{R_{10}} & -\frac{1}{R_7} & 0 \\ 0 & 0 & -\frac{1}{R_7} & \frac{1}{R_7}+\frac{1}{R_8}+\frac{1}{R_9} & -\frac{1}{R_9} \\ -\frac{1}{R_6} & 0 & 0 & -\frac{1}{R_9} & \frac{1}{R_6}+\frac{1}{R_9}+\frac{1}{R_{11}} \end{pmatrix} \begin{pmatrix} V_1 \\ V_2 \\ V_3 \\ V_4 \\ V_5 \end{pmatrix} = \begin{pmatrix} \frac{e_1}{R_1} \\ \frac{e_2}{R_2}-\frac{e_5}{R_5} \\ \frac{e_5}{R_5}+\frac{e_7}{R_7} \\ -\frac{e_7}{R_7} \\ 0 \end{pmatrix} \tag{5.88}$$

Without loss of generality, we can put $p_k = 1/R_k$, $k = 1, \ldots, 11$. We solve system (5.88) with tolerances 5, 10, 20, 23% from the nominal values of p_k. The quality of enclosures is assessed by using the sharpness measure. The results are presented in Tables 5.29 and 5.30. For 23% tolerance, the IAGSI was the only method that produced positive intervals.

The next table (Table 5.31) compares the results of the IAGSI method and the results from [182] for 10% tolerance. The third column of the table shows the overestimation O_w of the bounds produced by the method from [182] over the IAGSI bounds. The hull solution given as rational numbers (cf. [182]) is given in the last column of the table.

Example 5.54 Consider a linear AC electrical circuit (cf. Fig. 5.9). The parameters of the model have the following nominal values:

Table 5.29 Sharpness of outer enclosures for Example 5.53

Uncert. (%)	IAGSI	MRFPI	PIGSI	DM, PHBRC	GEM(2)
	Min–max	Min–max	Min–max	Min–max	Min–max
5	0.82–0.87	0.79–0.83	0.79–0.84	0.77–0.82	0.82–0.87
10	0.65–0.72	0.58–0.64	0.59–0.68	0.57–0.64	0.64–0.72
20	0.25–0.38	0.19–0.27	0.20–0.32	0.18–0.29	0.23–0.35
23	0.11–0.25	0.08–0.17	0.08–0.20	0.08–0.18	0.10–0.22

Table 5.30 Comparison of computational times for Example 5.53

Uncert. (%)	IAGSI	MRFPI	PIGSI	DM	PHBRC	GEM(2)
5	0.021(7)	0.011(1)	0.021(6)	0.006	0.005	0.034
10	0.025(8)	0.012(2)	0.028(8)	0.006	0.005	0.034
20	0.047(14)	0.016(3)	0.039(13)	0.006	0.005	0.034
23	0.054(17)	0.021(3)	0.044(16)	0.006	0.005	0.034

$$
\begin{aligned}
& e_1 = e_2 = 100\,\mathrm{V},\; e_5 = e_7 = 10\,\mathrm{V},\\
& Z_j = R_j + iX_j \in \mathbb{C},\; R_j = 100\,\Omega,\; X_j = \omega L_j - \frac{1}{\omega C_j},\; j = 1, \ldots, 11,\\
& \omega = 50,\; X_{1,2,5,7} = \omega L_{1,2,5,7} = 20,\; X_3 = \omega L_3 = 30,\\
& X_4 = -\frac{1}{\omega C_4} = -300,\; X_{10} = -\frac{1}{\omega C_{10}} = -400,\; X_{6,8,9,11} = 0.
\end{aligned}
$$

The worst-case tolerance analysis leads in this case to complex parametric interval linear system (cf. [105, 188])

$$
\begin{pmatrix}
\frac{1}{Z_1}+\frac{1}{Z_3}+\frac{1}{Z_6} & -\frac{1}{Z_3} & 0 & 0 & -\frac{1}{Z_6}\\
-\frac{1}{Z_3} & \frac{1}{Z_2}+\frac{1}{Z_3}+\frac{1}{Z_4}+\frac{1}{Z_5} & -\frac{1}{Z_4}-\frac{1}{Z_5} & 0 & 0\\
0 & -\frac{1}{Z_4}-\frac{1}{Z_5} & \frac{1}{Z_4}+\frac{1}{Z_5}+\frac{1}{Z_7}+\frac{1}{Z_{10}} & -\frac{1}{Z_7} & 0\\
0 & 0 & -\frac{1}{Z_7} & \frac{1}{Z_7}+\frac{1}{Z_8}+\frac{1}{Z_9} & -\frac{1}{Z_9}\\
-\frac{1}{Z_6} & 0 & 0 & -\frac{1}{Z_9} & \frac{1}{Z_6}+\frac{1}{Z_9}+\frac{1}{Z_{11}}
\end{pmatrix}
\begin{pmatrix} V_1\\ V_2\\ V_3\\ V_4\\ V_5 \end{pmatrix}
=
\begin{pmatrix} \frac{e_1}{Z_1}\\ \frac{e_2}{Z_2}-\frac{e_5}{Z_5}\\ \frac{e_5}{Z_5}+\frac{e_7}{Z_7}\\ -\frac{e_7}{Z_7}\\ 0 \end{pmatrix},
\tag{5.89}
$$

Table 5.31 Comparison of results of IAGSI method and results from [182] for Example 5.53 for 10% tolerance; hull solution is given in last column

Voltage	IAGSI	Result from [182]	O_w (%)	Hull
V_1	[55.33861, 66.87946]	[55.10036, 66.93172]	2	$[\frac{1075850}{19223}, \frac{1194650}{18143}]$
V_2	[50.82447, 62.70316]	[50.65584, 62.71315]	1	$[\frac{8839900}{171471}, \frac{1389700}{22541}]$
V_3	[32.17605, 43.46377]	[32.1721, 43.54983]	1	$[\frac{6048320}{181869}, \frac{2724960}{63757}]$
V_4	[13.36789, 22.70486]	[13.3486, 22.80113]	1	$[\frac{519228}{35897}, \frac{78012}{3521}]$
V_5	[21.26246, 31.41527]	[21.18106, 31.54621]	2	$[\frac{14016630}{629927}, \frac{53229110}{1725359}]$

Table 5.32 Comparison of enclosures for Example 5.54

Uncertainty (%)	IAGSI	MRFPI	PIGSI	DM, PHBRC	GEM(2)
	Min–max	Min–max	Min–max	Min–max	Min–max
5	0.64–0.85	0.61–0.81	0.60–0.82	0.59–0.80	0.64–0.85
10	0.28–0.67	0.25–0.61	0.24–0.61	0.23–0.58	0.27–0.66
15	0.00–0.44	0.00–0.36	0.00–0.38	0.00–0.35	0.00–0.42
20	0.00–0.10	0.00–0.08	0.00–0.08	0.00–0.07	0.00–0.09

Table 5.33 Comparison of computational times and number of iterations (in parentheses) for Example 5.54

Uncert. (%)	IAGSI	MRFPI	PIGSI	DM	PHBRC	GEM(2)
5	0.15(8)	0.08(2)	0.12(7)	0.032	0.029	0.274
10	0.21(11)	0.08(2)	0.15(10)	0.032	0.029	0.274
15	0.29(17)	0.10(3)	0.21(15)	0.032	0.029	0.274
20	0.44(31)	0.11(4)	0.39(27)	0.032	0.029	0.274

where

$$\begin{aligned}
Z_1 &= Z_2 = Z_5 = Z_7 = 100 + i20,\\
Z_3 &= 100 + i30,\\
Z_4 &= 100 - i300,\\
Z_6 &= Z_8 = Z_9 = Z_{11} = 100,\\
Z_{10} &= 100 - i400.
\end{aligned}$$

Similarly as in the previous example, we put $p_j = 1/Z_j$. We solve the system with tolerances $\pm 5\%$, $\pm 10\%$, $\pm 15\%$, and 20%. The results are presented in Tables 5.32, 5.33 and 5.34.

For 15% tolerance, IAGSI and GEM were the only methods that produced positive intervals for all real parts and negative intervals for all imaginary parts of the voltages. Whereas, for 20% tolerance, IAGSI and GEM methods were the only methods that produced the positive intervals for all real parts of voltages.

On the basis of all the obtained results, we can conclude that the considered methods are useful for solving various problems involving parametric interval linear systems, including practical problems of structural and electrical engineering. The IAGSI and GEM methods seem to be essentially the most efficient and, therefore, they are the most recommended.

Table 5.34 Results of the IAGSI method for Example 5.54

Uncertainty	10%
V_1	$[61.663286, 72.843351] + i[-9.127639, -4.009014]$
V_2	$[65.70333, 77.21764] + i[-10.99673, -4.183922]$
V_3	$[48.526391, 63.705541] + i[-16.62215, -6.34864]$
V_4	$[19.426686, 30.492173] + i[-9.238798, -3.970498]$
V_5	$[25.269545, 36.074428] + i[-6.415913, -2.379855]$
Uncertainty	20%
V_1	$[50.007343, 85.285188] + i[-17.864816, 4.362096]$
V_2	$[51.68396, 91.821452] + i[-22.885148, 7.039179]$
V_3	$[27.433663, 84.445243] + i[-34.964436, 11.113561]$
V_4	$[5.67336, 43.464689] + i[-19.250108, 5.663657]$
V_5	$[13.760536, 46.993275] + i[-13.743392, 4.657675]$

Chapter 6
Parametric Interval Linear Programming Problem

We shall now switch to the following *parametric interval linear programming* (PILP) problem (cf. [113]): given parametric objective function

$$f(x, p) = c^T(p)x, \tag{6.1}$$

where $c_i(p)$ $(i = 1, \ldots, n)$ are, in general, nonlinear functions of p, and constraint

$$A(p)x = b(p), \quad p \in \boldsymbol{p} \in \mathbb{IR}^K, \tag{6.2}$$

determine range

$$f^*(A(p), b(p), c(p), \boldsymbol{p}) = \left\{c^T(p)x(p) : \; A(p)x = b(p), p \in \boldsymbol{p}\right\}. \tag{6.3}$$

Obviously, endpoints $\underline{f}^*$ and $\overline{f}^*$ of range (6.3) can be determined by solving following two optimization problems

$$\underline{f}^* = \min\left\{f(x, p) : A(p)x = b(p), p \in \boldsymbol{p}\right\}, \tag{6.4a}$$

$$\overline{f}^* = \max\left\{f(x, p) : A(p)x = b(p), p \in \boldsymbol{p}\right\}, \tag{6.4b}$$

i.e., by solving optimization problem (5.5) with, respectively, $g(x, p) = f(x, p)$ and $g(x, p) = -f(x, p)$.

Parametric interval linear programming problem (6.1) and (6.2) is a generalization of the known *interval linear programming* (ILP) problem, where interval matrix $\boldsymbol{A}$ and interval vectors $\boldsymbol{b}$, $\boldsymbol{c}$ are involved (see, e.g., [79, 81, 82, 205, 216]). The PILP problem is more complex than the ILP problem, since the feasible set (see Sect. 4.3) has a very complex structure, in particular, it is not convex even in a single orthant.

I. Skalna, *Parametric Interval Algebraic Systems*, Studies in Computational Intelligence 766, https://doi.org/10.1007/978-3-319-75187-0_6

Moreover, the implicit dependence between x and p cause that the PILP problem is also more complex than classical parametric linear programming problems (see, e.g., [1, 22, 86, 100, 133, 204, 276]).

6.1 Iterative Method

In (Kolev and Skalna [118]) we proposed an iterative method for solving optimization problem (6.4a), which exploits p-solution of parametric interval linear system (6.2). Note that problem (6.4b) can be solved in, essentially, the same manner. The computational scheme of the method, which will be referred to as the MPILP method, is as follows. Starting with initial domain $\boldsymbol{p}^{(0)} = \boldsymbol{p}$, p-solution $\boldsymbol{x}(p)$ of system (6.2) is computed in domain $\boldsymbol{p}^{(v)}$, where v denotes current iteration. Substituting $\boldsymbol{x}(p)$ into (6.1) gives

$$f(x, p) = \textstyle\sum_{i=1}^{n} c_i(p)\boldsymbol{x}_i(p), \quad p \in \boldsymbol{p}^{(v)}. \tag{6.5}$$

Next, upper bound f^u on $\underline{f}^*$, is found in $\boldsymbol{p}^{(v)}$. It can be obtained from the IEH solution to the PILP problem [113], or by applying some local optimization method. Thus, the constraint equation is determined at each current v-th ($v \geqslant 0$) iteration corresponding to domain $\boldsymbol{p}^{(v)}$:

$$\sum_{i=1}^{n} c_i(p)\boldsymbol{x}_i(p) = f^u, \quad p \in \boldsymbol{p}^{(v)}. \tag{6.6}$$

A simple interval constraint satisfaction technique is now applied in order to reduce the current domain $\boldsymbol{p}^{(v)}$ to narrower domain $\boldsymbol{p}^{(v+1)}$. The progress in the domain reduction is measured by using distance $q\left(\boldsymbol{p}^{(v)}, \boldsymbol{p}^{(v+1)}\right)$. The distance between two interval vectors $\boldsymbol{x}, \boldsymbol{y} \in \mathbb{IR}^n$ is defined by [113]:

$$q(\boldsymbol{x}, \boldsymbol{y}) = \max\left\{\max_{i=1,\ldots,n} \left|\overline{x}_i - \overline{y}_i\right|, \max_{i=1,\ldots,n} \left|\underline{x}_i - \underline{y}_i\right|\right\} \tag{6.7}$$

If $q\left(\boldsymbol{p}^{(v)}, \boldsymbol{p}^{(v+1)}\right) \leqslant \varepsilon_q$, where ε_q is a prescribed threshold, the iterations are resumed. The iterative process continues until the width of the current domain becomes smaller than given threshold ε_p or else no progress in the domain reduction has been achieved (either $\boldsymbol{p}^{(v)} = \boldsymbol{p}^{(v+1)}$ or the reduction is negligible).

According to formula (5.4) (see Chap. 5) the p-solution of system (6.2) has the form $\boldsymbol{x}(p) = Lp + \boldsymbol{a}$, $p \in \boldsymbol{p}$. To simplify the presentation, only the special case of the PILP problem, when $c_i(p) = c_i \in \mathbb{R}$, $i = 1, \ldots, n$, is considered here. Let

$$m = c^T L, \tag{6.8}$$

$$\underline{\lambda} = \min\left\{m^T p \;:\; p \in \boldsymbol{p}\right\}, \tag{6.9a}$$

$$\overline{\lambda} = \max\left\{m^T p \;:\; p \in \boldsymbol{p}\right\}, \tag{6.9b}$$

$$\boldsymbol{g} = c^T \boldsymbol{a}, \tag{6.10}$$

Theorem 6.1 *Let* $\boldsymbol{f}^{(l)} = \underline{\lambda} + [\underline{g}, \overline{g}]$ *and* $\boldsymbol{f}^{(u)} = \overline{\lambda} + [\underline{g}, \overline{g}]$. *Then* $\underline{f}^* \in \boldsymbol{f}^{(l)}$ *and* $\overline{f}^* \in \boldsymbol{f}^{(u)}$.

Proof Let $p \in \boldsymbol{p}$. If $x(p)$ is a solution to parametric interval linear system (6.2), then $x(p) \in \boldsymbol{x}(p)$. Hence,

$$c^T x(p) \in c^T\left(Lp + \boldsymbol{a}\right) = c^T Lp + c^T \boldsymbol{a} = m^T p + \boldsymbol{g}.$$

So, we have

$$c^T x(p) \leqslant m^T p + \overline{g}, \tag{6.11a}$$
$$c^T x(p) \geqslant m^T p + \underline{g}. \tag{6.11b}$$

Since (6.11a) and (6.11b) hold for all $p \in \boldsymbol{p}$, hence

$$\underline{\lambda} + \underline{g} \leqslant \underline{f}^* \leqslant \underline{\lambda} + \overline{g},$$
$$\overline{\lambda} + \underline{g} \leqslant \overline{f}^* \leqslant \overline{\lambda} + \overline{g},$$

which proves the claim. □

So, the upper bound on $\underline{f}^*$ is given by

$$f^{(u,1)} = \underline{\lambda} + \overline{g}, \tag{6.12}$$

and the lower bound on $\overline{f}^*$ is given by

$$f^{(l,1)} = \overline{\lambda} + \underline{g}, \tag{6.13}$$

The method employing bounds (6.12) and (6.13), which will be referred to as MPILP.V1 method, is presented in Algorithm 12.

In order to simplify the computation, system (6.2) is first normalized and then the procedure P1 (see Algorithm 13) is used to contract the original parameters. The simplest possible interval constraint satisfaction is used here to narrow the current domain $\boldsymbol{p}^{(v)}$. It makes use solely of the constraint (6.14) and it contracts only one component of $\boldsymbol{p}^{(v)}$ at a time. A more sophisticated approach would be to propagate the constraint over the equations of the system (6.2).

The right-hand side of constraint equation (6.6) can also be determined by using a local optimization method. The method employing upper bound $f^{(u,2)}$ obtained

Algorithm 12 (MPILP.V1 method)

Input: $\boldsymbol{p}, \varepsilon_q, \varepsilon_p, v_{\max}$
$v = 0$;
$\boldsymbol{p}^{(v)} = \boldsymbol{p}$ // initial domain
repeat
$v = v + 1$
if $\left(\max\left\{\text{rad}\left(\boldsymbol{p}_i^{(v)}\right), i = 1, \ldots, K\right\} \leqslant \varepsilon_p\right)$ **then**
return $\boldsymbol{p}^*$ satisfying the accuracy conditions, containing the vector p^*, which yields the lower end-point $\underline{l}^*$ of $\boldsymbol{l}^*$ // "Success"
else
Normalize system $A(\boldsymbol{p}^{(v)})x = b(\boldsymbol{p}^{(v)})$
Calculate p-solution $\boldsymbol{x}(e)$ of system the normalized system
Calculate upper bound $f^{(u,1)}$ on $\underline{f}^*$ by using formulae (6.8)–(6.10) and (6.12)
Using $\boldsymbol{x}_i(e), i = 1, \ldots, n$, and $l^{(u,1)}$ construct the constraint equation

$$c^T a^c + m^T e + \boldsymbol{s} = l^{(u,1)}, \tag{6.14}$$

where $e \in \boldsymbol{e}$ and $\boldsymbol{s} = |c|^T a^{\Delta}[-1, 1]$
Apply the interval constraint satisfaction procedure P1 to obtain new (hopefully narrower) domain $\boldsymbol{p}^{(v+1)}$
if $(q\left(\boldsymbol{p}^{(v)}, \boldsymbol{p}^{(v+1)}\right) \leqslant \varepsilon_q)$ **then**
return only a crude two-sided bound on the lower end-point $\underline{f}^*$ of $\boldsymbol{f}^*$ has been found
end if
end if
until $(v > v_{\max})$

Algorithm 13 (Procedure P1)

Rewrite equation (6.14) in the form

$$m_k e_k = f^{(u,1)} - c^T a^c - \textstyle\sum_{j \neq k}^K m_j e_j - \boldsymbol{s},$$

where $k = \min\left\{j \mid j \in \{1, \ldots, K\} \text{ and } |m_j| = \max\limits_{i=1,\ldots,K} |m_i|\right\}$

Calculate $\boldsymbol{e}_k' = (\boldsymbol{b}/m_k) \cap \boldsymbol{e}_k$,
where $\boldsymbol{b} = d - \sum_{j \neq k}^K m_j e_j - \boldsymbol{s}$, $d = \sum_{i=1}^n |c_i| a_i^{\Delta} - \sum_{j=1}^K |m_j|$
if $(\boldsymbol{e}_k' = \boldsymbol{e}_k$ or reduction of $\text{wid}(\boldsymbol{e}')$ is negligible) **then**
return no progress in the reduction of the current domain $\boldsymbol{p}^{(v)}$ has been achieved
end if
Three cases can be distinguished:
Case A. if $\underline{e}_k' = \underline{e}_k = -1$, then also $\underline{p}_k' = \underline{p}_k$. Upper endpoint $\overline{p}_k'$ is found from the relation $(\overline{p}_k' - \underline{p}_k)/(\overline{p}_k - \underline{p}_k) = (\overline{e}_k' - \underline{e}_k)/(\overline{e}_k - \underline{e}_k)$, which leads to $\overline{p}_k' = \underline{p}_k + p_k^{\Delta}(\overline{e}_k' + 1)$
Case B. If $\overline{e}_k' = \overline{e}_k = 1$, then $\overline{p}_k' = \overline{p}_k$. Using similar reasoning as in Case A, we obtain $\underline{p}_k' = \overline{p}_k - p_k^{\Delta}(1 - \underline{e}_k')$
Case C. If $\boldsymbol{e}_k'$ is strictly included in $\boldsymbol{e}_k$, i.e., $\underline{e} < \underline{e}_k'$ and $\overline{e}_k' < \overline{e}_k$, then both endpoints of $\boldsymbol{p}_k'$ must be determined. On account of Case A, lower endpoint $\underline{p}_k'$ is given by $\underline{p}_k' = \underline{p}_k + p_k^{\Delta}(\underline{e}_k' + 1)$. Similarly, $\overline{p}_k' = \underline{p}_k + p_k^{\Delta}(\overline{e}_k' + 1)$.
return New reduced-width domain $\boldsymbol{p}' = [\underline{p}_k', \overline{p}_k']$

Table 6.1 Asymptotic time complexity of methods MPILP.V1, MPILP.V2, MPILP.V3: M - number of iterations of the M2 method, n - size of PILP problem, K - number of parameters, κ - number of iterations of M1 method, g - number of generations in P2 procedure

Method	Asymptotic time complexity
MPILP.V1	$\mathcal{O}(M(n^3 + \kappa n^2 K^2))$
MPILP.V2	$\mathcal{O}(M(gn^3 + \kappa n^2 K^2))$
MPILP.V3	$\mathcal{O}(M(n^3 + \kappa n^2 K^2))$

using ES method from Sect. 5.3.2 will be referred to as MPILP.V2 method. Another way to obtain upper bound $f^{(u,3)}$ on $\underline{f}^*$ is as follows. Let $\tilde{p}$ have components

$$\tilde{p}_j = \begin{cases} \underline{p}_j, & m_j > 0, \\ \overline{p}_j, & \text{otherwise}. \end{cases} \tag{6.15}$$

Find solution $\tilde{x}$ corresponding to the system $A(\tilde{p})x = b(\tilde{p})$ and compute $l^{(u,3)} = c^T\tilde{x}$. The method employing bound $f^{(u,3)}$ will be referred to as MPILP.V3 method. The asymptotic time complexity of the proposed methods is given in Table 6.1.

To compare the performance of the variants of the MPILP method, the special case of problem (6.1) and (6.2) is considered here, where

$$c = (1, 1, 1)^T. \tag{6.16}$$

Example 6.2 The aim of this two-dimensional example is to graphically illustrate the PILP problem. The constraint equation is given by

$$A(p) = \begin{pmatrix} p_1 & p_2 + 1 \\ p_2 + 1 & -3p_1 \end{pmatrix}, \ b(p) = \begin{pmatrix} 2p_1 \\ 1 \end{pmatrix}. \tag{6.17}$$

Two cases are considered: (A) $p_1, p_2 \in [0.3, 0.55]$, (B) $p_1, p_2 \in [0.2, 0.65]$. The feasible sets for both cases and the points at which the minimum and maximum values are attained, are presented in Fig. 6.1.

Example 6.3 In this example the constraint equation is given by

$$A(p) = \begin{pmatrix} p_1 & p_2 + 1 & -p_3 \\ p_2 + 1 & -3 & p_1 \\ 2 - p_3 & 4p_2 + 1 & 1 \end{pmatrix}, \ b(p) = \begin{pmatrix} 2p_1 \\ p_3 - 1 \\ -1 \end{pmatrix}, \tag{6.18}$$

and the involved parameter vectors (boxes) of variable width, which depends on a parameter ρ, have the following form

$$p(\rho) = p^c + \rho \cdot p^{\Delta}[-1, 1], \tag{6.19}$$

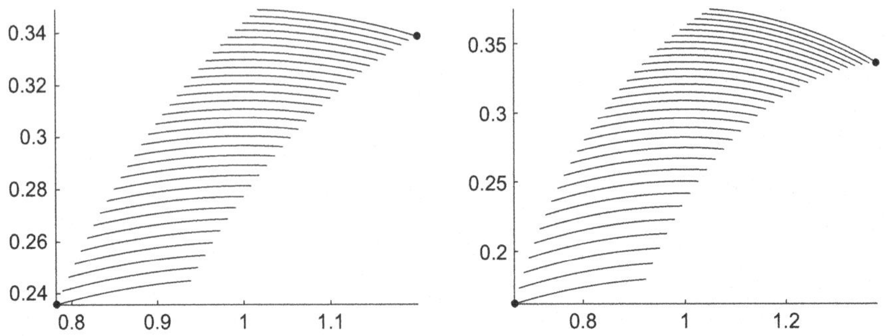

Fig. 6.1 Feasible set defined by the constraint (6.17) and points at which minimum and maximum values are attained; case A (left), case B (right)

Table 6.2 Hull solution to PILP problem 6.2 obtained using MPILP.V1, MPILP.V2 and MPILP.V3 methods, $\rho = 0.1$

Method	$\underline{l}^*$	#iter	$\overline{l}^*$	#iter	t [s]
MPILP.V1	−1.336397202	11	−1.152408631	11	0.064
MPILP.V2	−1.336397202	8	−1.152408631	10	0.223
MPILP.V3	−1.336397202	8	−1.152408631	10	0.058

where $p^c = (0.5, 0.5, 0.5)$, $p^\Delta = (0.5, 0.5, 0.5)$. The results for $\rho = 0.1$ are presented in Table 6.2 (the MPILP.V2 method is run with $N = 10$, $T = 10$, $b = 6$, $N_x = 0.1N$, $\pi_c = 0.2$, $\pi_m = 0.7$). Table 6.2 shows the hull solution, the number of iterations (#iter) for each endpoint and the overall computational time given in seconds.

It can be seen that the MPILP.V3 method is the best out of the three methods with respect to time complexity. However, time complexity is only one of several factors that can be used to determine the efficiency of a method. It can also be measured by using the so-called *radius of applicability* [112], which is defined as follows:

$$r_a(M) = \sup\{\rho \mid M \text{ is applicable to } Pr(\boldsymbol{p}(\rho)) \text{ for } \boldsymbol{p}(\rho)\}, \tag{6.20}$$

where $Pr(\boldsymbol{p}(\rho))$ denotes an interval analysis problem defined for a given interval vector $\boldsymbol{p}(\rho) = p^c + \rho[-p^\Delta, p^\Delta]$. If $r_a(M_1) < r_a(M_2)$, then M_2 is numerically more efficient, since M_1 fails to solve the problem earlier (for an interval vector of smaller width) than the M_2 method.

The radius of applicability can be estimated by increasing ρ by Δq. The value of r_a(MPILP.V1), r_a(MPILP.V2) and r_a(MPILP.V3) are given in the third column of Table 6.3. Near the "critical" value, the increment of ρ was chosen to be $\Delta\rho = 0.001$. As we can see, r_a(MPILP.V1) $\ll r_a$(MPILP.V2) $= r_a$(MPILP.V3), which means, taking into account the previous results, that the MPILP.V3 method is the most efficient among the three considered variants.

Table 6.3 Radius of applicability of the MPILP.V1, MPILP.V2 and MPILP.V3 methods

Method	$\rho = r_a$	$\underline{f}^*$	#iter	$\overline{f}^*$	#iter
MPILP.V1	0.180	−1.424531295	37	−1.092690377	42
MPILP.V2	0.269	−1.533073244	14	−1.035389370	46
MPILP.V3	0.269	−1.533073244	14	−1.035389370	45

Table 6.4 Data on the MPILP.V4 method, $\rho = 0.269$

$\underline{f}^*$	τ	#iter	$\overline{f}^*$	τ	#iter
−1.533073244	0.74	8	−1.035389370	0.71	13

The convergence of the MPILP.V3 method can be further improved, i.e., smaller number of iterations can be achieved, by taking $f^{(u,4)} < f^{(u,3)}$, since $f^{(u,4)}$ is more contracting. A good candidate is

$$f^{(u,4)}(\tau) = \underline{f} + \tau\left(f^{(u,3)} - \underline{f}\right), \tag{6.21}$$

where $\underline{f}$ is a lower bound of OI solution to the PILP problem and $\tau \in [0, 1]$. The lower bound on $\underline{f}^*$ in this case will be

$$f^{(l,4)}(\tau) = \overline{f} + \tau\left(f^{(l,3)} - \overline{f}\right). \tag{6.22}$$

The method employing the bounds (6.21) and (6.22) will be referred to as MPILP.V4 method (it is worth to note that the asymptotic time complexity of the MPILP.V4 method is $\mathcal{O}(M(n^3 + \kappa n^2 K^2))$. Now the question is how much less can/should be $f^{(u,4)}$ than $f^{(u,3)}$, so that the intersection $\boldsymbol{b}/m_k \cap \boldsymbol{e}_k$ (see Procedure 1) was not empty. Using several experiments, it has been established that computing $\underline{f}^*$ takes the least number of iterations when τ ranges from 0.74–0.78, whereas the minimum number of iterations for $\overline{f}^*$ is achieved when τ ranges from 0.68–0.71. Data on the MPILP.V4 method are given in Table 6.4. It can be seen that the number of iterations has been significantly decreased. Another advantage of using the MPILP.V4 method is that is has larger radius of applicability. With $\tau = 0.74$ for $\underline{l}^*$ and $\tau = 0.71$ for $\overline{l}^*$, the radius of applicability r_a(MPILP.V4) $= 0.333$ (see Table 6.5). Moreover, using the MPILP.V4 method, the lower bound of $\boldsymbol{f}^*$ can be determined up to $\rho = 0.595$. So, the radius of applicability for the partial problem, i.e. for the problem of computing $\underline{l}^*$ is r_a(MPILP.V4) $= 0.595$. The results are presented in Table 6.6. The last column shows the percentage by which the obtained result overestimates the hull solution.

Future research will concentrate on enhancing the numerical efficiency and the applicability of the proposed approach. Possible ways are to use some computationally more efficient methods yielding $\boldsymbol{x}(p)$ or to use appropriate quadratic interval enclosures to approximate higher-order parametric functions. Also, a more sophis-

Table 6.5 Data on the MPILP.V4 method, $\rho = r_a(\text{MPILP.V4}) = 0.333$

$\underline{f}^*$	τ	#iter	$\overline{f}^*$	τ	#iter
−1.618373240	0.74	8	−1.0001701722	0.71	13

Table 6.6 Data on the MPILP.V4 method, $\rho = 0.595$, $\tau = 0.74$

$\underline{f}^*$	#iter	$\overline{f}^*$	#iter
−2.039682970	28	−0.79733463331	8

ticated interval constraint satisfaction technique (e.g., involving all equations of the LIP system) would improve the convergence of the proposed method.

Summary

Starting in the mid-nineties, there has been a growing interest in methods for solving problems involving interval data. This great interest was caused by the fact that interval methods produce validated results in the presence of both bounded data uncertainty and rounding errors (which are inherent in numerical computing). However, the bounds produced by many early interval methods were rather crude and, thus, of little practical use. The main reason for this deficiency was the dependency problem (Sect. 1.3) of interval arithmetic. Therefore, significant research in the field of interval analysis was directed towards the development of methods that were able to take data dependency into account. An important part of this research (which is of main interest in this book) is related to the problem of solving parametric interval linear systems (Sect. 4.2); i.e., systems whose coefficients are functions of the parameters varying within prescribed intervals.

The present monograph not only summarizes the author's scientific research, reported in a series of papers [245–257], but also presents new results not published elsewhere. Our *algorithmic contribution* is the development of a unified framework and methods (Chap. 5) based on revised affine arithmetic (Sect. 2.2) for solving parametric interval linear systems in a rigorous manner. The choice of the revised affine arithmetic as a tool for range analysis was dictated by several reasons. First of all, revised affine forms can be straightforwardly used to model both affine-linear and nonlinear dependencies. Moreover, revised affine arithmetic preserves all features of standard affine arithmetic (i.e., it automatically keeps track of rounding and truncation errors; it also keeps track of correlations between input data); whereas, it is computationally more efficient since the length of the revised affine forms remains unchanged in the course of the same computation. Further, revised affine arithmetic (like standard affine arithmetic) keeps track of the correlations between intermediate results, thus reducing the so-called wrapping effect. Finally, the solutions obtained using revised affine arithmetic have the form of p-solutions, which enables us to obtain both an inner estimate of the hull solution and outer interval enclosure for the parametric solution set. This is an important feature since the comparison of inner and outer bounds enables us to assess the quality of the latter. Moreover, p-solutions can be useful in solving various other problems involving parametric dependency.

I. Skalna, *Parametric Interval Algebraic Systems*, Studies in Computational Intelligence 766, https://doi.org/10.1007/978-3-319-75187-0

The numerical experiments performed in Chap. 5, which include practical problems of structural and electrical engineering, indicate that the developed framework is very useful for solving parametric interval linear systems with both affine-linear and nonlinear dependencies. The recently developed Interval-affine Gauss-Seidel Iteration method (Sect. 5.2.2) seems to be one of the best methods for solving such systems. All of the methods yield guaranteed bounds, which is of great importance for practical applications. An important feature of the proposed approach is that it can be used to solve both real and complex systems (in the latter case, the approach so far is limited to systems with affine-linear dependencies) as well as rectangular systems. Our *theoretical contribution* concerns the development of an arithmetic of revised affine forms (Chap. 2) and a comprehensive overview of the theory of parametric interval matrices (Chap. 3). The presented theory might be useful not only for solving parametric interval linear systems but also for solving many other problems related to parametric interval matrices.

Below, we give an overview of our main results and present possible directions of future research.

- In Chap. 2, arithmetic operations on revised affine forms is developed. In particular, a new efficient algorithm for computing the Chebyshev minimum-error multiplication of revised affine forms is proposed. The algorithm has linear (in the number of noise symbols) asymptotic time complexity, which is optimal. Since division is performed using multiplication, the division also gains linear time complexity. The lower time complexity of these operations can extend the scope of the practical applications of revised affine forms. Possible directions for further research in the field of range analysis are the following:

(a) further improve arithmetic of revised affine forms,
(b) employ parallel techniques to decrease the computational time of affine computation,
(c) combine revised affine forms with other techniques for range bounding,
(d) develop arithmetic of complex revised affine forms,
(e) develop efficient tools for handling the dependency problem using higher order forms.

- In Chap. 3, the theory of parametric interval matrices is presented. A large part of this chapter is concerned with investigating the basic properties of parametric interval matrices such as regularity, strong regularity (which are of great importance for the problem of solving parametric interval linear systems), and Hurwitz and Schur stability. It has been proven that checking the regularity and stability of a parametric interval matrix are NP-hard problems. In order to circumvent the exponential complexity, several verifiable sufficient conditions for checking these properties are proposed. The presented results extend the known result for interval matrices. Possible directions of future research in this area include the following issues:

(a) develop more-powerful conditions for checking the discussed properties of parametric interval matrices,
(b) develop respective algorithms for checking these properties,
(c) develop an algorithm for computing the radius of regularity/stability,
(d) other interesting topics such as eigenvalue range determination.

- In Chap. 5, various methods for solving parametric interval linear systems are elaborated. They include approximate methods, methods producing the p-solution, and methods for computing the hull solution. In particular, a modification of Rump's fixed-point iteration is presented (Skalna [245]). This modification extends the applicability of Rump's method and guarantees narrower enclosures. Next, the Direct Method (Skalna [248]) is described. It is then generalized to solve parametric interval linear systems with a multiple right-hand side. Furthermore, the Bauer-Skeel and Hansen-Bliek-Rohn methods are presented. The residual correction of the right-hand vector (which aims to improve the bounds) is also discussed. It has been proven that the Direct Method and Bauer-Skeel method are equivalent if a system is preconditioned with the inverse of the midpoint matrix. It has also been shown that, both in the classical and parametric cases, the Hansen-Bliek-Rohn method is never worse than the Bauer-Skeel method. Next, two modifications of the Monotonicity Approach (Skalna [250]) are proposed. The first one was suggested by Kolev; the second (which is based on the p-solution) is a new result that has not been published as of yet. Next, Kolev's linear and quadratic parametric methods are briefly presented. The Parametrized Direct method of Kolev is then described in more details. Also, an iterative parametric method that was recently developed by Skalna and Hladík (Skalna and Hladík [259]) is presented, and some general results on the convergence of iterative methods are proven. Further, the generalization of the Expansion method is suggested. Finally, methods for computing the exact interval solution to parametric interval linear systems are described: the Interval Global Optimization and Evolutionary methods. Further research in this area will be focused on the following issues:

(a) develop methods that do not require strong regularity of system matrix,
(b) improve the efficiency of existing methods by using parallel techniques, for example,
(c) employ p-solution to solve a wider class of parametric interval linear systems,
(d) develop methods for computing bounds for AE solution sets,
(e) develop methods that combine various range bounding techniques,
(f) extend the applicability of the methods to control problems, for example.

- In Chap. 6, a new method for solving a special case of a parametric interval linear programming problem is proposed. The method can be further improved by:

(a) elaborating a more complex but (hopefully) more-efficient constraint propagation technique,
(b) finding a better way of selecting the upper bound for the computed minimum,
(c) using some more-sophisticated local optimization method,

(d) employing a more-efficient method for computing the p-solution,
(e) a challenging problem is to address a more-realistic PLP setting where matrix $A(p)$ is rectangular or there are inequality constraints,
(f) the method can also be extended to solve other linear programming problems.

References

1. S. Aggarwal, Parametric linear fractional functionals programming. Metrika **12**(1), 106–114 (1967). https://doi.org/10.1007/BF02613489
2. R. Akhmerov, Interval-Affine Gaussian algorithm for constrained systems. Reliab. Comput. **11**, 323–341 (2005)
3. G. Alefeld, On the approximation of the range of values by interval expressions. Computing **44**, 273–278 (1990)
4. G. Alefeld, J. Herzberger, Über die Verbesserung von Schranken für die Lösung bei linearen Gleichungssystemen. Angew. Informatik **3**, 107–112 (1971)
5. G. Alefeld, J. Herzberger, *Introduction to Interval Computations*, Computer science and applied mathematics (Academic Press Inc, New York, 1983). Transl. by J. Rokne from the original German 'Einführung In Die Intervallrechnung'
6. G. Alefeld, V. Kreinovich, G. Mayer, On the shape of the symmetric, persymmetric and skew-symmetric solution set. SIAM J. Matrix Anal. Appl. **18**(3), 693–705 (1997). https://doi.org/10.1137/S0895479896297069
7. G. Alefeld, V. Kreinovich, G. Mayer, The shape of the solution set for systems of interval linear equations with dependent coefficients. Math. Nachr. **192**, 23–36 (1998)
8. G. Alefeld, R. Lohner, On higher order centered forms. Computing **35**, 177–184 (1985)
9. G. Alefeld, G. Mayer, The cholesky method for interval data. Linear Algebra Appl. **194**, 161–182 (1993)
10. G. Alefeld, J. Rokne, On the evaluation of rational functions in interval arithmetic. SIAM J. Numer. Anal. **18**(5), 862–870 (1981). https://doi.org/10.1137/0718059
11. Archimedes, in *The Works of Archimedes*, dover edition 1953 edn., On the measurement of the circle, ed. by T.L. Heath (Cambridge University Press, Cambridge, 1897)
12. E. Bareiss, Sylvester's identity and multistep integer-preserving gaussian elimination. Math. Comput. **103**, 565–578 (1968)
13. B. Barmish, New tools for robustness analysis, in *Proceedings of the 27th Conference on Decision and Control* (Austin, TX, 1988), pp. 1–6
14. F. Bauer, Zusammenfassender Bericht. Genauigkeitsfragen bei der Lösung linearer Gleichungssysteme. ZAMM - J. Appl. Math. Mech./Zeitschrift für Angewandte Mathematik und Mechanik **46**(7), 409–421 (1966). https://doi.org/10.1002/zamm.19660460702
15. E. Baumann, Optimal centered forms. BIT **28**, 80–87 (1988)
16. H. Beeck, Charakterisierung der lösungsmenge von intervallgleichungssystemen. Z. Angew. Math. Mech. **53**, 181–182 (1973)
17. H. Beeck, Zur Problematik Der Hüllenbestimmung Von Intervallgleichungssystemen. Lect. Notes Comput. Sci. **29**, 150–159 (1975)

I. Skalna, *Parametric Interval Algebraic Systems*, Studies in Computational Intelligence 766, https://doi.org/10.1007/978-3-319-75187-0

18. M. Berz, K. Makino, Rigorous global search using taylor models. Symb. Numer. Comput. **11–19**, 2009 (2009)
19. S. Bialas, A necessary and sufficient condition for the stability of interval matrices. Int. J. Control **37**, 717–722 (1983)
20. C. Bliek, Computer Methods for Design Automation. Dissertation, Massachsetts Institute of Technology, Dept. of Ocean Engineering, Cambridge, 1992
21. J. Buckley, Y. Qu, On using α-cuts to evaluate fuzzy equations. Fuzzy Sets Syst. **38**(3), 309–312 (1990). https://doi.org/10.1016/0165-0114(90)90204-J
22. A. Cambini, S. Schaible, C. Sodini, Parametric linear fractional programming for an unbounded feasible region. J. Global Optim. **3**(2), 157–169 (1993). https://doi.org/10.1007/BF01096736
23. G. Caplat, Symbolic preprocessing in interval function computing, in *Symbolic and Algebraic Computation., Lecture Notes in Computer Science*, vol. 72, ed. by G. Goos, J. Hartmanis (1979), pp. 369–382
24. O. Caprani, K. Madsen, Iterative methods for interval inclusion of fixed points. BIT Numer. Math. **18**(1), 42–51 (1978). https://doi.org/10.1007/BF01947742
25. O. Caprani, K. Madsen, Mean value forms in interval analysis. Computing **25**(2), 147–154 (1980)
26. G. Cargo, O. Shiska, The Bernstein form of a polynomial. J. Res. Nat. Bur. Stand. **70B**, 79–81 (1966)
27. F. Chernousko, Optimal guaranteed estimates of indeterminacies using ellipsoids, Parts 13. Izv. Akad. Nauk SSSR, Tekh. Kibern. (1980). 3,311;4,311;5,511
28. J. Comba, J. Stolfi, Affine arithmetic and its applications to computer graphics, in *Proceedings of SIBGRAPI'93 VI Simpósio Brasileiro de Computação Gráfica e Processamento de Imagens* (Recife, BR, 1993), pp. 9–18
29. G. Corliss, Industrial applications of interval techniques. Contrib. Comput. Arith. Self-Validating Numer. Methods IMACS Ann. Comput. Appl. Math. **7**, 91–113 (1990)
30. G. Corliss, C. Foley, R. Kearfott, Formulation for reliable analysis of structural frames. Reliab. Comput. **13**(2), 125–147 (2007). https://doi.org/10.1007/s11155-006-9027-0
31. H. Cornelius, R. Lohner, Computing the range of values of real functions with accuracy higher than second order. Computing **33**, 331–347 (1984)
32. T. Csendes, D. Ratz, Subdivision direction selection in interval methods for global optimization. SIAM J. Numer. Anal. **34**(3), 922–938 (1997). https://doi.org/10.1137/S0036142995281528
33. L. de Figueiredo, J. Stolfi, *Self-Validated Numerical Methods and Applications*, Brazilian mathematics colloquium monographs. IMPA/CNPq (Rio de Janeiro, Brazil, 1997)
34. L. de Figueiredo, J. Stolfi, An introduction to affine arithmetic. TEMA Tends Appl. Comput. Math. **4**(3), 297–312 (2003)
35. L. de Figueiredo, J. Stolfi, Affine arithmetic: concepts and applications. Numer. Algorithms **37**(1–4), 147–158 (2004)
36. D. Degrauwe, G. Lombaert, G.D. Roeck, Improving interval analysis in finite element calculations by means of affine arithmetic. Comput. Struct. **88**(3–4), 247–254 (2010)
37. M. Dehghani-Madiseh, M. Dehghan, Parametric AE-solution sets to the parametric linear systems with multiple right-hand sides and parametric matrix equation $A(p)X = B(p)$. Numer. Algorithms **73**(1), 245–279 (2016). https://doi.org/10.1007/s11075-015-0094-3
38. N. Delanoue, L. Jaulin, B. Cottenceau, An algorithm for computing a neighborhood included in the attraction domain of an asymptotically stable point. Commun. Nonlinear Sci. Numer. Simul. **21**(1–3), 181–189 (2015)
39. J. Demmel, The componentwise distance to the nearest singular matrix. SIAM J. Matrix Anal. Appl. **13**(1), 10–19 (1992). https://doi.org/10.1137/0613003
40. A. Dreyer, Interval analysis of analog circuits with component tolerances. Ph.D. thesis, TU Kaiserslautern, Aachen, Germany, 2005
41. J. Duda, I. Skalna, differential evolution applied to large scale parametric interval linear systems. Lect. Notes Comput. Sci. **7116**, 206–213 (2012). https://doi.org/10.1007/978-3-642-29843-1_23

42. J. Duda, I. Skalna, Heterogeneous multi-agent evolutionary system for solving parametric interval linear systems. Lect. Notes Comput. Sci. **7782**, 465–472 (2013). https://doi.org/10.1007/978-3-642-36803-5_35
43. J. Duda, I. Skalna, GPU acceleration of metaheuristics solving large scale parametric interval algebraic systems. Lect. Notes Comput. Sci. **8385**, 591–599 (2014). https://doi.org/10.1007/978-3-642-55195-6_56
44. T. Dzetkulič, Rigorous computation with function enclosures in Chebyshev basis (2012)
45. H. El-Owny, Parametric linear system of equations, whose elements are nonlinear functions, in *12th GAMM - IMACS International Symposion on Scientific Computing, Computer Arithmetic and Validated Numerics* (2006)
46. H. El-Owny, *Solving Parametric Interval Linear Systems: Parametric Linear Systems, Interval Arithmetic C-XSC* (Lambert, 2012)
47. H. El-Owny, Improved iterative method for solving parametric linear systems. Int. J. Adv. Res. Technol. **2**(8), 296–300 (2013)
48. H. El-Owny, Outer interval solution of linear systems with parametric interval data. Int. J. Sci. Eng. Res. **4**(9), 1432–1436 (2013)
49. W. Enger, Interval ray tracing - a divide and conquer strategy for realistic computer graphics. Vis. Comput. **9**, 91–104 (1992)
50. Y. Fang, Optimal bicentered form. Reliab. Comput. **9**, 291–302 (2003)
51. M. Fiedler, *Special Matrices and Their Applications in Numerical Mathematics* (SNTL Publishing House, Prague, 1986)
52. I.P.-. W.G., for Interval Arithmetic, in *Introduction to the IEEE 1788-2015 Standard for Interval Arithmetic*
53. G. Forsythe, Pitfalls in computation, or why a math book isn't enough. Am. Math. Mon. **77**(9), 931–956 (1970). https://doi.org/10.2307/2318109
54. A. Gaganov, Computational complexity of the range of the polynomial in several variables. Master's thesis, Leningrad University, Math. Department, 1981
55. E. Gardenes, A. Trepat, Fundamentals of SIGLA, an interval computing system over the completed set of intervals. Computing **24**, 161–179 (1980)
56. M. Garey, D. Johnson, *Computers and Intractability: A Guide to the Theory of NP-Completeness* (Freeman, San Francisco, 1979)
57. J. Garloff, Convergent bounds for the range of multivariate polynomials, in *Interval Mathematics 1985, vol. 212*, Lecture Notes in Computer Science, ed. by K. Nickel (Springer, Berlin, 1986), pp. 37–56
58. J. Garloff, The bernstein expansion and its applications. J. Am. Romanian Acad. **25–27**(80–85), 2000–2003 (2003)
59. J. Garloff, C. Jansson, A. Smith, Inclusion isotonicity of convex-concave extensions for polynomials based on Bernstein expansion. Computing **70**, 111–119 (2003)
60. J. Garloff, E. Popova, A. Smith, Solving linear systems with polynomial parameter dependency with application to the verified solution of problems in structural mechanics, in *Optimization, Simulation, and Control*, Springer optimization and its applications, ed. by A. Chinchuluun, P.M. Pardalos, R. Enkhbat, E. Pistikopoulos (Springer, Berlin, 2013), pp. 301–318
61. J. Garloff, A. Schabert, A. Smith, Bounds on the range of multivariate rational functions. PAMM Proc. Appl. Math. Mech. **12**, 649–650 (2012)
62. J. Garloff, A. Schabert, A. Smith, Bounds on the range of multivariate rational functions. Proc. Appl. Math. Mech. (PAMM) **12**(1), 649–650 (2012)
63. J. Garloff, A. Smith, Special issue on the use of Bernstein polynomials in reliable computing: a centennial anniversary. Reliab. Comput. **00**, 17 (2012)
64. A. Goldsztejn, A right-preconditioning process for the formal-algebraic approach to inner and outer estimation of AE-solution sets. Reliab. Comput. **11**(6), 443–478 (2005)
65. A. Goldsztejn, G. Chabert, On the approximation of linear ae-solution sets. in *12th GAMM - IMACS International Symposium on Scientific Computing, Computer Arithmetic and Validated Numerics (SCAN 2006)* (2006), pp. 18–18. http://web.emn.fr/x-info/gchabe08/goldsztejn_chabert_scan06.pdf

66. G. Golub, W. Kahan, Calculating the singular values and pseudo-inverse of a matrix. J. Soc. Ind. Appl. Math. Ser. B, Numer. Anal. **2**(2), 205–224 (1965)
67. G. Golub, H. van der Vorst, Eigenvalue computation in the 20th century. J. Comput. Appl. Math. **123**(1), 35–65 (2000). Numerical Analysis 2000. Vol. III: Linear Algebra
68. B. Gross, Verification of asymptotic stability for interval matrices and applications in control theory, *Scientific Computing with Authomatic Result Verification* (1993), pp. 357–395
69. E. Hansen, Interval arithmetic in matrix computations, Part I. J. Soc. Ind. Appl. Math. Ser. B, Numer. Anal. **2**(2), 308–320 (1965). https://doi.org/10.2307/2949786
70. E. Hansen, On solving systems of equations using interval arithmetic. Math. Comput. **22**, 374–384 (1968)
71. E. Hansen, The centered form, in *Topics in Interval Analysis*, ed. by E. Hansen (Oxford University Press, London, 1969), pp. 102–106
72. E. Hansen, A generalized interval arithmetic, in *Interval Mathematics*, Lecture notes in computer science, ed. by K. Nickel (Springer, New York, 1975), pp. 7–18. https://doi.org/10.1007/3-540-07170-9_2
73. E. Hansen, Global optimization using interval analysis - the one-dimensional case. J. Optim. Theory Appl. **29**, 331–344 (1979)
74. E. Hansen, *Global Optimization Using Interval Analysis* (Marcel Dekker, New York, 1992)
75. E. Hansen, S. Sengupta, Bounding solutions of systems of equations using interval analysis. BIT (Nordisk tidskrift for informationsbehandling) **21**, 203–211 (1981)
76. E. Hansen, R. Smith, A Computer Program for Solving a System of Linear Equations and Matrix Inversion with Automatic Error Bounding Using Interval Arithmetic. Technical report LMSC 4-22-66-3, Lockheed Missiles and Space Co., Palo Alto, CA, 1966
77. D. Hertz, The extreme eigenvalues and stability of real symmetric interval matrices. IEEE Trans. Autom. Control **37**(4), 532–535 (1992). https://doi.org/10.1109/9.126593
78. M. Hladík, Description of symmetric and skew-symmetric solution set. SIAM J. Matrix Anal. Appl. **30**(2), 509–521 (2008)
79. M. Hladík, Optimal value range in interval linear programming. Fuzzy Optim. Decis. Mak. **8**(3), 283–294 (2009). https://doi.org/10.1007/s10700-009-9060-7
80. M. Hladík, Enclosures for the solution set of parametric interval linear systems. Int. J. Appl. Math. Comput. Sci. **22**(3), 561–574 (2012). https://doi.org/10.2478/v10006-012-0043-4
81. M. Hladík, An interval linear programming contractor. in *Proceedings of 30th International Conference on Mathematical Methods in Economics 2012 (Part I.)* (Karviná, Czech Republic, 2012. Silesian University in Opava, School of Business Administration in Karviná), pp. 284–289
82. M. Hladík, How to determine basis stability in interval linear programming. Optim. Lett. **8**(1), 375389 (2014)
83. M. Hladík, Optimal preconditioning for the interval parametric Gauss–Seidel method. Lect. Notes Comput. Sci. **9553**, 116–125 (2016). https://doi.org/10.1007/978-3-319-31769-4_10
84. M. Hladík, Positive semidefiniteness and positive definiteness of a linear parametric interval matrix (2017). arXiv:1704.05782
85. M. Hladík, E. Popova, Maximal inner boxes in parametric AE-solution sets with linear shape. Appl. Math. Comput. **270**, 606–619 (2015). https://doi.org/10.1016/j.amc.2015.08.003
86. A. Holder, *Parametric LP Analysis* (John Wiley & Sons, Inc., New York, 2010). https://doi.org/10.1002/9780470400531.eorms0643
87. H. Hong, V. Stahl, Bernstein form is inclusion monotone. Computing **55**(1), 43–53 (1995)
88. J. Horáček, M. Hladík, Computing enclosures of overdetermined interval linear systems. Reliab. Comput. **19**(2), 142–155 (2013)
89. C. Jacobi, Über ein leichtes verfahren, die in der theorie der säkularstörungen vorkommenden gleichungen numerisch aufzulösen. Crelle's J. (in German) **30**, 51–94 (1846)
90. C. Jansson, Interval linear systems with symmetric matrices, skew-symmetric matrices and dependencies in the right hand side. Computing **46**(3), 265–274 (1991)
91. L. Jaulin, M. Kieffer, O. Didrit, E. Walter, *Applied Interval Analysis with Examples in Parameter and State Estimation, Robust Control and Robotics* (Springer, New York, 2001)

92. Q. Lin, J.G. Rokne, Interval approximation of higher order to the ranges of functions. Comput. Math. Appl. **31**(7), 101–109 (1996)
93. W. Kahan, A logarithm too clever by half (2004). http://www.cs.berkeley.edu/~wkahan/LOG10HAF.TXT
94. L.W. Kantorowicz, O niektórych nowych podejściach do metod obliczeniowych i opracowaniu badań (in russian). Sybirskie Czasopismo Matematyczne **3**(5), 701–709 (1962)
95. E. Kaucher, Algebraische Erweiterungen der Intervallrechnung unter Erhaltung Ordnungs- und Verbandsstrukturen. Comput. Suppl. **1**(1), 65–79 (1977)
96. E. Kaucher, Interval Analysis In the Extended Interval Space IR. Computing. Supplementum **2**, 33–49 (1980)
97. R. Kearfott, *Rigorous global search: continous problems (Kluwer Academic Publishers Group, Norwell (USA, and Dordrecht* (The Netherlands, MA, 1996)
98. R. Kearfott, K. Du, The cluster problem in multivariate global optimization. J. Glob. Optim. **5**, 253–265 (1994)
99. R. Kearfott, V. Kreinovich (eds.), *Applications of Interval Computations* (Kluwer Academic Publishers Group, Dordrecht, 1996). The Netherlands
100. R. Khalilpour and I. Karimi. Parametric optimization with uncertainty on the left hand side of linear programs. *Computers & Chemical Engineering*, 60(Supplement C):31–40, 2014. https://doi.org/10.1016/j.compchemeng.2013.08.005
101. G. Klir, B. Yuan, *Fuzzy Sets and Fuzzy Logic: Theory and Applications* (Prentice Hall Inc., USA, 1995)
102. L. Kolev, *Interval Methods for Circuit Analysis* (Word Scientific Ltd., London, 1993)
103. L. Kolev, A new method for global solution of systems of nonlinear equations. Reliab. Comput. **4**(1), 1–21 (1998)
104. L. Kolev, Automatic Computation of a Linear Interval Enclosure. Reliable Computing **7**(1), 17–28 (2001)
105. L. Kolev, Worst-case tolerance analysis of linear DC and AC electric circuits. IEEE Trnas. Circuits Syst. I, Fund. Theory Appl. **49**(12), 1693–1702 (2002)
106. L. Kolev, An improved interval linearization for solving non-linear problems. Numer. Algorithms **37**(1–4), 213–224 (2004)
107. L. Kolev, A method for outer interval solution of linear parametric systems. Reliab. Comput. **10**(3), 227–239 (2004)
108. L. Kolev, Solving linear systems whose elements are nonlinear functions of intervals. Numer. Algorithms **37**(1–4), 213–224 (2004)
109. L. Kolev, Improvement of a direct method for outer solution of linear parametric systems. Reliab. Comput. **12**(3), 193–202 (2006)
110. L. Kolev, Optimal multiplication of G-intervals. Reliab. Comput. **13**(5), 399–408 (2007)
111. L. Kolev, A method for determining the regularity radius of interval matrices. Reliab. Comput. **16**, 1–26 (2011)
112. L. Kolev, Componentwise determination of the interval hull solution for linear interval parameter systems. Reliab. Comput. **20**(1), 1–24 (2014)
113. L. Kolev, Parametrized solution of linear interval parametric systems. Appl. Math. Comput. **246**(11), 229–246 (2014)
114. L. Kolev, Regularity radius and real eigenvalue range. Appl. Math. Comput. **233**, 404–412 (2014). https://doi.org/10.1016/j.amc.2014.01.113
115. L. Kolev, A Direct Method for Determining a p-Solution of Linear Parametric Systems. Journal of Applied & Computational Mathematics **5**(1), 1–5 (2016)
116. L. Kolev. Iterative algorithms for determining a p-solution of linear interval parametric systems. In *Advanced Aspects of Theoretical Electrical Engineering*, Sofia, Bulgaria, 15.09-16.09 2016
117. L. Kolev, A new class of iterative interval methods for solving linear parametric systems. Reliab. Comput. **22**, 26–46 (2016)
118. L. Kolev, I. Skalna, Exact solution to a parametric linear programming problem. Numer. Algorithms (2017). (accepted for publication)

119. R. Krawczk, Interval extensions and interval iterations. Computing **24**, 119–129 (1980)
120. R. Krawczyk, Newton-Algorithmen zur Bestimmung von Nullstellen mit Fehlerschranken. Computing **4**(3), 187–201 (1969). https://doi.org/10.1007/BF02234767
121. R. Krawczyk, Fehlerabschätzung bei Linearer Optimierung. Lect. Notes Comput. Sci. **29**, 215–222 (1975)
122. Computing Centered forms and interval operators. **34**(3), 243–259 (1985). https://doi.org/10.1007/BF02253320
123. R. Krawczyk, A. Neumaier, Interval slopes for rational functions and associated centered forms. SIAM J. Numer. Anal. **22**(3), 604–616 (1985). https://doi.org/10.2307/2157086
124. V. Kreinovich, For interval computations, if absolute-accuracy optimization is NP-hard, then so is relative-accuracy optimization (1999)
125. V. Kreinovich, Range estimation is NP-Hard for ϵ^2 accuracy and feasible for $\epsilon^{2-\delta}$. Reliab. Comput. **8**(6), 481–491 (2002)
126. V. Kreinovich, A. Lakeyev, S. Noskov, Optimal solution to interval linear systems is intractable (NP-hard). Interval Comput. **1**, 6–14 (1993)
127. V. Kreinovich, A. Lakeyev, J. Rohn, P. Kahl, *Computational Complexity and Feasibility of Data Processing and Interval Computations*, 1st edn. (Springer, Dordrecht, 1998). https://doi.org/10.1007/978-1-4757-2793-7
128. U. Kulisch, R. Lohner, A. Facius, *Perspectives on Enclosure Methods* (Springer Science and Business Media, Berlin, 2012)
129. O. Lanford, Computer-assisted proofs in analysis. Phys. A: Stat. Mech. Appl. **124**(1), 465–470 (1984). https://doi.org/10.1016/0378-4371(84)90262-0
130. Q. Lin, J. Rokne, Methods for bounding the range of a polynomial. J. Comput. Appl. Math. **58**, 193–199 (1995)
131. W. Lodwick. Constrained Interval Arithmetic. Technical Report CCM Report 138, University of Colorado at Denver, Denver, USA, February 1999
132. J. Lunze, *Robust Multivariable Feedback Control* (Prentice-Hall, Englewood Cliffs, 1989)
133. M. Černý, M. Hladík, Inverse optimization: towards the optimal parameter set of inverse LP with interval coefficients. Central Eur. J. Oper. Res. **24**(3), 747–762 (2016). https://doi.org/10.1007/s10100-015-0402-y
134. M. Mansour, Robust stability of interval matrices. in *Proceedings of the 28th IEEE Conference on Decision and Control*, vol. 1 (1989), pp. 46–51
135. R. Martin, H. Shou, I. Voiculescu, A. Bowyer, G. Wang, Comparison of interval methods for plotting algebraic curves. Compu. Aided Geom. Des. **19**(7), 553–587 (2002)
136. J. Matthews, R. Broadwater, L. Long, The application of interval mathematics to utility economics analysis. IEEE Trans. Power Syst. **5**(1), 177–181 (1990)
137. G. Mayer, An Oettli-Prager-like theorem for the symmetric solution set and for related solution sets. SIAM J. Matrix Anal. Appl. **33**(3), 979–999 (2012). https://doi.org/10.1137/120867895
138. G. Mayer, Three short descriptions of the symmetric and of the skew-symmetric solution set. Linear Algebra Appl. **475**, 73–79 (2015). https://doi.org/10.1016/j.laa.2015.02.003
139. F. Messine, New affine forms in interval branch and bound algorithms. Technical report R2I 99-02, Université de Pau et des Pays de l'Adour (UPPA), France, 1999
140. F. Messine, Extentions of affine arithmetic: application to unconstrained global optimization. J. Univers. Comput. Sci. **8**(11), 992–1015 (2002)
141. C. Meyer, *Matrix Analysis and Applied Linear Algebra* (SIAM, Philadelphia, 2000)
142. S. Miyajima, On the improvement of the division of the affine arithmetic. Master's thesis, Kashiwagi Laboratory, Waseda University, Japan, 2000
143. S. Miyajima, M. Kashiwagi, A dividing method utilizing the best multiplication in affine arithmetic. IEICE Electron. Express **1**(7), 176–181 (2004)
144. S. Miyajima, T. Miyata, M. Kashiwagi, On the best multiplication of the affine arithmetic. IEICE Trans. **J86–A(2)**, 150–159 (2003)
145. R. Moore, Interval arithmetic and automatic error analysis in digital computing, Ph.D. thesis, Stanford University, 1962
146. R. Moore, *Interval Anal.* (N. J. Prentice-Hall Inc, New York, 1966)

147. R. Moore, *Methods and Applications of Interval Analysis*, Society for industrial and applied mathematics (Philadelphia, USA, 1979)
148. R. Moore, *Reliability in Computing: The Role of Interval Methods in Scientific Computing* (Academic Press, New York, 1988)
149. R. Moore, On reducing overestimation of ranges of multinomials without splitting boxes (2006). http://www.mat.univie.ac.at/~neum/ms/Moore-ranges.pdf
150. R. Moore, R. Kearfott, J. Cloud, *Introduction to Interval Analysis*, Society for industrial and applied mathematics (Philadelphia, USA, 2009)
151. R. Moore, C. Yang, Interval Analysis I. Technical document LMSD-285875, Lockheed Missiles and Space Division, Sunnyvale, CA, USA, 1959
152. S. Mudur, P. Koparkar, Interval methods for processing geometric objects. IEEE Comput. Graph. Appl. **4**(2), 7–17 (1984)
153. R. Mullen, R. Muhanna, Bounds of structural response for all possible loading combinations. J. Struct. Eng. 98–106 (1999)
154. A. Narkawicz, J. Garloff, A. Smith, C. Muñoz, Bounding the range of a rational function over a box. Reliab. Comput. **17**, 34–39 (2012)
155. P. Nataraj, K. Kotecha, Higher order convergence for multidimensional functions with a new taylor-bernstein form as inclusion function. Reliab. Comput. **9**(3), 185–203 (2003)
156. P. Nataraj, K. Kotecha, Global optimization with higher-order inclusion function forms Part 1: a combined TaylorBernstein form. Reliab. Comput. **10**(1), 27–44 (2004)
157. P. Nataraj, S. Tharewal, A computational approach to existence verification and construction of robust QFT controllers, in *Proceedings*, NSF workshop on reliable engineering computing, Savannah, GA, 2004, ed. by R. Muhanna, R. Mullen
158. A. Neumaier, Tolerance analysis with interval arithmetic. Freiburger Intervall-Berichte **86**(9), 5–19 (1986)
159. A. Neumaier, Overestimation in linear interval equations. SIAM Journal of Numerical Analysis **24**(1), 207–214 (1987). https://doi.org/10.2307/2157396
160. A. Neumaier, *Interval Methods for Systems of Equations*, Encyclopedia of mathematics and its applications (Cambridge University Press, UK, 1990)
161. A. Neumaier, Taylor forms - use and limits. Reliab. Comput. **9**(1), 43–79 (2003)
162. A. Neumaier, Improving interval enclosures (2009). http://www.mat.univie.ac.at/~neum/ms/encl.pdf
163. A. Neumaier, A. Pownuk, Linear systems with large uncertainties, with applications to truss structures. Reliab. Comput. **13**(2), 149–172 (2007)
164. A. Niemirowski, Several NP-hard problems arising in robust stability analysis. Math. Control Signals Syst. **6**(2), 99–105 (1993). https://doi.org/10.1007/BF01211741
165. W. Oettli, On the solution set of a linear system with inaccurate coefficients. J. Soc. Ind. Appl. Math. Ser. B Numer. Anal. **2**(1), 115–118 (1965). https://doi.org/10.1137/0702009
166. W. Oettli, W. Prager, Compatibility of Approximate Solution of Linear Equations with Given Error Bounds for Coefficients and Right-Hand Sides. Numerische Mathematik **6**, 405–409 (1964)
167. W. Oettli, W. Prager, J. Wilkinson, Admissible solutions of linear systems with not sharply defined coefficients. J. Soc. Ind. Appl. Math. Ser. B Numer. Anal. **2**(2), 291–299 (1965). https://doi.org/10.1137/0702023
168. K. Okumura, An application of interval operation to electric network analysis. Bulletin of the Japan Society for Industrial and Applied Mathematics **3**(2), 115–127 (1993). https://doi.org/10.11540/bjsiam.3.2_115
169. A. Ostrowski, Über die Determinanten mit überwiegender hauptdiagonale. Comment. Math. Helv. **10**, 69–96 (1937)
170. S. Parsons, Interval algebras and order of magnitude reasoning, in *Applications of Artificial Intelligence in Engineering VI*, ed. by G. Rzevski, R. Adey (Elsevier Applied Science, UK, 1991), pp. 945–961
171. S. Poljak, J. Rohn, Checking robust nonsingularity is NP-hard. Math. Control Signals Syst. **6**(1), 1–9 (1993)

172. E. Popova, Computer algebra in solving linear systems with imprecise dependent data. Reliab. Comput. Interval Algebra (2000)
173. E. Popova, *On the Solution of Parametrised Linear Systems* (Springer, US, 2001), pp. 127–138. https://doi.org/10.1007/978-1-4757-6484-0_11
174. E. Popova, Generalization of a parametric fixed-point iteration. in *Proceedings in Applied Mathematics and Mechanics* (2004), pp. 680–681
175. E. Popova, Strong regularity of parametric interval matrix, in *33rd Spring Conference of the Union of Bulgarian Mathematicians, Mathematics and Education in Mathematics*, ed. by I. Dimovski, K. Bankov, P. Boivalenkov, G. Ganchev, M. Dobreva, A. Eskenazi, C. Lozanov (Borovets, Bulgaria, April 1–4 2004), pp. 446–451
176. E. Popova, Computer-assisted proofs in solving linear parametric problems, *12th GAMM/IMACS International Symposium on Scientific Computing, Computer Arithmetic and Validated Numerics, SCAN 2006* (Duisburg, Germany, 2006), pp. 35–35
177. E. Popova, Improved solution enclosures for over- and underdetermined interval linear systems (Springer, Heidelberg, 2006), pp. 305–312. https://doi.org/10.1007/11666806_34
178. E. Popova, Solving linear systems whose input data are rational functions of interval parameters. Lect. Notes Comput. Sci. **4310**, 345–352 (2007). https://doi.org/10.1007/978-3-540-70942-8_41
179. E. Popova, Explicit characterization of a class of parametric solution sets. Comptes rendus de lAcadémie Bulgare des Sciences **62**(10), 1207–1216 (2009)
180. E. Popova, Explicit description of AE solution sets for parametric linear systems. SIAM J. Matrix Anal. Appl. **33**(4), 1172–1189 (2012). https://doi.org/10.1137/120870359
181. E. Popova, Inner estimation of the parametric tolerable solution set. Comput. Math. Appl. **66**(9), 1655–1665 (2013). BioMath 2012
182. E. Popova, Improved enclosure for some parametric solution sets with linear shape. Comput. Math. Appl. **68**(9), 994–1005 (2014). https://doi.org/10.1016/j.camwa.2014.04.005
183. E. Popova, On the unbounded parametric tolerable solution set. Numer. Algorithms **69**(1), 169–182 (2015). https://doi.org/10.1007/s11075-014-9888-y
184. E. Popova, Outer bounds for the parametric controllable solution set with linear shape. Lect. Notes Comput. Sci. **9553**, 138–147 (2016)
185. E. Popova, Enclosing the solution set of parametric interval matrix equation $A(p)X = B(p)$. Numer. Algorithms (2017). https://doi.org/10.1007/s11075-017-0382-1
186. E. Popova, M. Hladík, Outer enclosures to the parametric AE solution set. Soft Comput. **17**(8), 1403–1414 (2013). https://doi.org/10.1007/s00500-013-1011-0
187. E. Popova, R. Iankov, Z. Bonev, Bounding the Response of Mechanical Structures with Uncertainties in all the Parameters, in *Proceedings of the NSF Workshop on Reliable Engineering Computing (REC'06)*, ed. by I .R .L. Muhanna, R .L. Mullen (Savannah, Georgia USA, Feb. 22–24 2006), pp. 245–265
188. E. Popova, L. Kolev, W. Krämer, A solver for complex-valued parametric linear systems. Serdica J. Comput. **4**(1), 00 (2010)
189. E. Popova, W. Krämer, Inner and outer bounds for parametric linear systems. Comput. Appl. Math. **199**(2), 310–316 (2007)
190. E. Popova, W. Krämer, Visualizing parametric solution sets. BIT Numer. Math. **48**(1), 95–115 (2008)
191. E. Popova, W. Krämer, Characterization of AE Solution Sets to a Class of Parametric Linear Systems. Comptes Rendus de L'Académie Bulgare des Sciences **64**(3), 325–332 (2011)
192. A. Pownuk, Efficient method of solution of large scale engineering problems with interval parameters, in *NSF workshop on Reliable Engineering Computing* (Savannah, Georgia, USA, September 15-17 2004
193. H. Ratschek, Centered forms. Numer. Anal. **17**(5), 656–662 (1977)
194. H. Ratschek, J. Rokne, About the centered form. SIAM J. Numer. Anal. **17**(3), 333–337 (1980). https://doi.org/10.1137/0717027
195. H. Ratschek, J. Rokne, Optimality of the centered form, in *Interval Mathematics*, Springer, 1980, ed. by K. Nickel, pp. 499–508

196. H. Ratschek, J. Rokne, Optimality of the centered form for polynomials. J. Approx. Theory **32**, 151–159 (1981)
197. H. Ratschek, J. Rokne, *New Computer Methods for Global Optimization* (Halsted Press, USA, 1988)
198. H. Ratschek, G. Schröder, Centered forms for functions in several variables. J. Math. Anal. Appl. **82**, 534–552 (1981)
199. D. Ratz, On extended interval arithmetic and inclusion isotonicity (1996)
200. N. Revol, P. Théveny, Numerical reproducibility and parallel computations: issues for interval algorithms. IEEE Trans. Comput. **63**(8), 1915–1924 (2014). https://doi.org/10.1109/TC.2014.2322593
201. G. Rex, J. Rohn, A note on checking regularity of interval matrices. Linear Multilinear Algebra **39**(3), 259–262 (1995)
202. G. Rex, J. Rohn, Sufficient conditions for regularity and singularity of interval matrices. SIAM J. Matrix Anal. Appl. **20**(2), 437–445 (1999)
203. F. Ris, Interval analysis and applications to linear algebra, Ph.D. thesis, University of Oxford, Oxford, 1972
204. K. Ritter, A method for solving maximum-problems with a nonconcave quadratic objective function. Zeitschrift für Wahrscheinlichkeitstheorie und Verwandte Gebiete **4**(4), 340–351 (1966). https://doi.org/10.1007/BF00539118
205. J. Rohn, Strong solvability of interval linear programming problems. Computing **26**(1), 79–82 (1981). https://doi.org/10.1007/BF02243426
206. J. Rohn, Interval Linear Systems. Freiburger Intervall-Ber. 84/7, Universität Freiburg, Freiburg, Germany, 1984
207. J. Rohn, Solving Interval Linear Systems, in *Collection of Scientific Papers Honouring Prof. Dr. K. Nickel on Occasion of His 60th Birthday, Part II*, ed. by J.G. et al. (Inst. F. Angew. Math., University Freiburg I, Br, 1984), pp. 419–432
208. J. Rohn, Systems of linear interval equations. Linear Algebra Appl. **126**, 39–78 (1989)
209. J. Rohn, Cheap and tight bounds: the recent result by E. Hansen can be made more efficient. Interval Comput. **4**, 13–21 (1993)
210. J. Rohn, Inverse interval matrix. SIAM J. Numer. Anal. **30**(3), 864–870 (1993)
211. J. Rohn, NP-hardness results for linear algebraic problems with interval data, in *Topics in Validated Computations Studies in Computational Mathematics*, ed. by J. Herzberger (1994), pp. 463–472
212. J. Rohn, Positive definiteness and stability of interval matrices. SIAM J. Matrix Anal. Appl. **15**(1), 175–184 (1994)
213. J. Rohn, Checking properties of interval matrices. Technical report 686, Institute of Computer Science, Academy of Sciences of the Czech Republic, Prague, 1996. http://www.cs.cas.cz/~rohn/publist/92.ps
214. J. Rohn, Linear Interval Equations: Computing Enclosures with Bounded Relative Overestimation is NP-Hard, in *Applications of interval computations: Papers presented at an international workshop in El Paso, Texas, February 23–25, 1995*, Applied optimization, ed. by R. Kearfott, V. Kreinovich (Kluwer Academic Publishers Group, Dordrecht, The Netherlands, 1996), pp. 81–90
215. J. Rohn, A method for handling dependent data in interval linear systems. Technical report 911, Institute of Computer Science, Academy of Sciences of the Czech Republic, Prague, 2004
216. J. Rohn, Interval linear programming (Springer, Boston, 2006), pp. 79–100. https://doi.org/10.1007/0-387-32698-7_3
217. J. Rohn, An improvement of the Bauer-Skeel bounds. Technical report V-1065, Institute of Computer Science, Academy of Sciences of the Czech Republic, Czech, 2010
218. J. Rohn, A characterization of strong regularity of interval matrices. Electron. J. Linear Algebra **20**(2010)
219. J. Rohn, Explicit inverse of an interval matrix with unit midpoint. Electron. J. Linear Algebra **22**, 138–150 (2011). https://doi.org/10.13001/1081-3810.1430

220. J. Rohn, V. Kreinovich, Computing exact componentwise bounds on solutions of linear systems with interval data is NP-Hard. SIAM J. Matrix Anal. Appl. **16**(2), 415–420 (1995)
221. J. Rohn, G. Rex, Interval P-matrices. SIAM J. Matrix Anal. Appl. **17**, 1020–1024 (1996)
222. J. Rokne, Optimal computation of the Bernstein algorithm for the bound of an interval polynomial. Computing **28**, 239–246 (1982)
223. J. Rokne, A low complexity explicit rational centered form. Computing **34**(3), 261–263 (1985)
224. J. Rokne, Low Complexity k-dimensional Centered Forms. Computing **37**, 247–253 (1986)
225. S. Rump, Kleine Fehlerschranken Bei Matrixproblemen, Dissertation, Universität Karlsruhe, 1980
226. S. Rump, Solving Algebraic Problems with High Accuracy. in *Proceedings of the Symposium on A New Approach to Scientific Computation* (Academic Press Professional, Inc., USA, 1983), pp. 51–120
227. S. Rump, *New Results on Verified Inclusions* (Springer, Heidelberg, 1986), pp. 31–69. https://doi.org/10.1007/3-540-16798-6_4
228. S. Rump, On the solution of interval linear systems. Computing **47**(3), 337–353 (1992). https://doi.org/10.1007/BF02320201
229. S. Rump, Verification methods for dense and sparse systems of equations, in *Topics in validated computations: proceedings of IMACS-GAMM International Workshop on Validated Computation, Oldenburg, Germany, 30 August–3 September 1993*, Studies in computational mathematics, ed. by J. Herzberger (Elsevier, Amsterdam, The Netherlands, 1993), pp. 63–136
230. S. Rump, The distance between regularity and strong regularity. Math. Res. **90**, 105–117 (1996)
231. S. Rump, A note on epsilon-inflation. Reliab. Comput. **4**(4), 371–375 (1998). https://doi.org/10.1023/A:1024419816707
232. S. Rump, *INTLAB — Interval Laboratory* (Kluwer Academic Publishers, Dordrecht, 1999), pp. 77–104. https://doi.org/10.1007/978-94-017-1247-7_7
233. S. Rump, Interval operations in rounding to nearest (2007)
234. S. Rump, Verification methods: rigorous results using floating-point arithmetic. Acta Numerica **19**, 287–449 (2010)
235. S. Rump, M. Kashiwagi, Implementation and improvements of affine arithmetic. Nonlinear Theory Appl. IEICE **6**(3), 341–359 (2015)
236. S. Rump, E. Kaucher, Small bounds for the solution of systems of linear equations. Comput. Suppl. **2**, 157–164 (1980)
237. A. Schrijver, *Theory of Linear and Integer Programming* (Wiley, New York, 1986)
238. D. Schwartz, S. Chen, Order of magnitude reasoning for qualitative matrix structural analysis. in *Proceedings of the Fifth International Conference on Computing in Civil and Building Engineering* (1993)
239. S. Sergey, Controllable solution set to interval static systems. Appl. Math. Comput. **86**(2), 185–196 (1997). https://doi.org/10.1016/S0096-3003(96)00181-6
240. I. Sharaya, S. Shary, Tolerable solution set for interval linear systems with constraints on coefficients. Reliab. Comput. **15**, 345–357 (2011)
241. S. Shary, On controlled solution set of interval algebraic systems. Interval Comput. **4**(6), 66–75 (1992)
242. S. Shary, Solving the tolerance problem for linear interval equations. Interval Comput. **2**, 6–26 (1994)
243. S. Shary, Solving the linear interval tolerance problem. Math. Comput. Simul. **39**, 53–85 (1995)
244. S. Shary, A new technique in systems analysis under interval uncertainty and ambiguity. Reliab. Comput. **8**(5), 321–418 (2002). https://doi.org/10.1023/A:1020505620702
245. I. Skalna, Metody rozwiązywania sparametryzowanych układów równań liniowych, *inżynieria wiedzy i systemy ekspertowe, Z. Bubnicki, [Knowledge Engineering and Expert Systems, in Polish]*, vol. 1(2000), pp. 92–102
246. I. Skalna, Methods for solving systems of linear equations of structure mechanics with interval parameters. Comput. Assist. Mech. Eng. Sci. **10**(3), 281–293 (2003)

247. I. Skalna, Zastosowanie metod algebry przedziałowej w jakościowej analizie układów mechanicznych, Ph.D. thesis, Politechnika Śląska, 2003. (in polish)
248. I. Skalna, A method for outer interval solution of parametrized systems of linear interval equations. Reliab. Comput. **12**(2), 107–120 (2006)
249. I. Skalna, Evolutionary optimization method for approximating the solution set hull of parametric linear systems. Lecture Notes Comput. Sci. **4310**, 361–368 (2007)
250. I. Skalna, On checking the monotonicity of parametric interval solution. Lecture Notes Comput. Sci. **4967**, 1400–1409 (2008)
251. I. Skalna, Direct method for solving parametric interval linear systems with non-affine dependencies. Lecture Notes Comput. Sci. **6068**, 485–494 (2010)
252. I. Skalna, A global optimization method for solving parametric linear systems whose input data are rational functions of interval parameters. Lecture Notes Comput. Sci. **6068**, 475–484 (2010)
253. I. Skalna, Enclosure for the solution set of parametric linear systems with non-affine dependencies(Springer, Heidelberg, 2012), pp. 513–522. https://doi.org/10.1007/978-3-642-31500-8_53
254. I. Skalna, Algorithm for min-range multiplication of affine forms. Numer. Algorithms **63**(4), 601–614 (2013)
255. I. Skalna, Strong regularity of parametric interval matrices. Linear Multilinear Algebra **65**(12), 2472–2482 (2017)
256. I. Skalna, J. Duda, A comparison of metaheurisitics for the problem of solving parametric interval linear systems. Lecture Notes Comput. Sci. **6046**, 305–312 (2011)
257. I. Skalna, A study on vectorisation and paralellisation of the monotonicity approach, Lecturer notes computer science (Springer International Publishing, Cham, 2016), pp. 455–463. https://doi.org/10.1007/978-3-319-32152-3_42
258. I. Skalna, M. Hladík, A new algorithm for Chebyshev minimum-error multiplication of reduced affine forms. Numer. Algorithms **76**(4), 1131–1152 (2017). https://doi.org/10.1007/s11075-017-0300-6
259. I. Skalna, M. Hladík, A new method for computing a p-solution to parametric interval linear systems with affine-linear and nonlinear dependencies. BIT Numer. Math. 1–28 (2017). https://doi.org/10.1007/s10543-017-0679-4
260. R. Skeel, Scaling for numerical stability in gaussian elimination. J. ACM **26**, 494–526 (1979)
261. S. Skelboe, Computation of rational interval functions. BIT **17**(1), 87–95 (1975)
262. A. Smith, Fast construction of constant bound functions for sparse polynomials. J. Glob. Optim. **43**(2–3), 445–458 (2009)
263. http://www.ic.unicamp.br/~stolfi/export/projects/affine-arith/welcome.html
264. P. Struss, Mathematical aspects of qualitative reasoning. Int. J. Artif. Intell. Eng. **3**(3), 156–169 (1988). lipiec. First Annual Workshop on Qualitative Physics
265. R. Sun, Y. Zhang, A. Cui, A refined affine approximation method of multiplication for range analysis in word-length optimization. EURASIP J. Adv. Signal Process. **1**, 2014 (2014)
266. T. Sunaga, Theory of an interval analysis and its application to numerical analysis. RAAG MEMOIRS **2**, 29–46 (1958)
267. L. Trefethen, D.B. III, *Numerical Linear Algebra* (Society for Industrial and Applied Mathematics, Philadelphia, 1997)
268. L. Tsoukalas, R. Uhrig, *Fuzzy and Neural Approaches in Engineering*, J. Wesley Hines (Wiley, New York, 1997)
269. A.K.I. Vályi, *Ellipsoidal Calculus for Estimation and Control* (Birkhäuser, Boston, 1997)
270. S. Vavasis, *Nonlinear Optimization: Complexity Issues* (Oxford University, New York, 1991)
271. X. Vu, D. Sam-Haroud, B. Faltings, A Generic Scheme for Combining Multiple Inclusion Representations in Numerical Constraint Propagation. Technical report IC200439, Swiss Federal Institute of Technology in Lausanne (EPFL), Switzerland, 2004
272. Y. Wang, Interpretable interval constraint solvers in semantic tolerance analysis. Comput.-Aided Des. Appl. **5**(5), 654–666 (2008). https://doi.org/10.3722/cadaps.2008.654-666

273. Y. Wang, Solving interval constraints in computer-aided design, in *Proceedings NSF Workshop on Reliable Engineering Computing (REC'04)*, ed. by R. Muhanna, R. Mullen (Savannah, GA, Sept. 15–17 2004), pp. 251–267
274. M. Warmus, Calculus of approximations. Bull. Acad. Polon. Sci. **4**(5), 253–259 (1956)
275. M. Warmus, Approximations and inequalities in the calculus of approximations. classification of approximate numbers. Bull. Acad. Polon. Sci. Ser. Sci. Math. Astronom. Phys. **9**, 241–245 (1961)
276. M. Wittmann-Hohlbein, E. Pistikopoulos, On the global solution of multi-parametric mixed integer linear programming problems. J. Glob. Optim. **57**(1), 51–73 (2013). https://doi.org/10.1007/s10898-012-9895-2
277. L. Zadeh, G. Klir, I. Turksen, P. Wang, *Advances in Fuzzy Theory and Technology* (Paul P, Wang, 1995)
278. B. Zalewski, R. Muhanna, R. Mullen, Bounding the response of mechanical structures with uncertainties in all the parameters, in *Proceedings of the NSF Workshop on Reliable Engineering Computing (REC'06), Feb. 22–24*, ed. by R. Muhanna, R. Mullen (Savannah, Georgia USA, 2006), pp. 439–456
279. M. Zettler, J. Garloff, Robustness analysis of polynomials with polynomial parameter dependency using Bernstein expansion. IEEE Trans. Autom. Control **43**(3) (1998)
280. L. Zhang, Y. Zhang, W. Zhou, Tradeoff between approximation accuracy and complexity for range analysis using affine arithmetic. J. Signal Process. Syst. **61**(3), 279–291 (2010)
281. G. Zielke, Report on test matrices for generalized inverses. Computing **36**(1), 105–162 (1986). https://doi.org/10.1007/BF02238196
282. G. Zielke, V. Drygalla, Genaue Lösung linearer Gleichungssyteme. Mitteilungen der GAMM **26**(1/2), 7–107 (2003)
283. M. Zimmer, W.K.E. Popova, Solvers for the verified solution of parametric linear systems. Computing **94**(2), 109–123 (2012). https://doi.org/10.1007/s00607-011-0170-z
284. Software Problem Led to System Failure at Dhahran, Saudi Arabia. IMTEC-92-26: Published: Feb 4, 1992. Publicly Released: Feb 27, 1992
285. http://extreme.adorio-research.org/download/mvf/html/node3.html
286. http://www.mat.univie.ac.at/~neum/glopt/moretest/
287. https://www.sfu.ca/~ssurjano/optimization.html

Index

A
absolute value, 4
affine
  ~ approximation, 29
    minimum-error ≈, 29
    minimum-range ≈, 29
  ~ form, 26
  ~ transformation, 58, 94
  approximation ~, 29
arithmetic
  affine ~, 25
  interval ~, 1, 6
  Kaucher interval ~, 18
  reduced affine ~, 34
  revised affine ~, 25, 34

B
Bernstein
  ~ coefficient, 15
  ~ expansion, 15
  ~ polynomial, 15

D
dependency, 8
  affine-linear ~, 57
  nonlinear ~, 57
distance
  Hausdorff ~, 5

E
eigenvalue
  real maximum magnitude ~, 72
ellipsoid calculus, 25
enclosure
  interval ~, 7
error
  accumulative ~, 33, 34
  roundout ~, 22
evaluation
  interval ~, 7
excess width, 10
extension
  interval ~, 9
  Lipschitz interval ~, 10
  natural interval ~, 7
  united ~, 7

F
form
  affine ~, 26
  Bernstein ~, 15
  bicentered ~, 13
  centered ~, 10
  generalized centered ~, 12
  mean value ~, 11
  mid-rad ~, 4
  optimal centered ~, 13
  reduced affine ~, 33
  revised affine ~, 33, 34
  Taylor-Bernstein ~, 16

H
hull, 4

I
inclusion
  ~ isotonicity, 5
  ~ monotonicity, 5

I. Skalna, *Parametric Interval Algebraic Systems*, Studies in Computational Intelligence 766, https://doi.org/10.1007/978-3-319-75187-0

interval ∼, 10
inclusion principle, 5
interval, 3
∼ intersection, 5
generalized ∼, 34
Hansen generalized ∼, 25
improper ∼, 19
interval linear system, 85
complex parametric ∼, 96
over-determined parametric ∼, 97
parametric ∼, 87
under-determined parametric ∼, 97
interval matrix, 51
inverse ∼, 53
parametric ∼, 57
iteration
fixed-point ∼, 101
Gauss–Seidel ∼, 128
Interval-affine Gauss–Seidel ∼, 124
Jacobi ∼, 128
Krawczyk ∼, 128
Parametric Gauss–Seidel ∼, 105

K
Krawczyk
operator ∼, 101

L
Lipschitz
condition ∼, 10

M
magnitude, 4
matrix
H-∼, 56
M-∼, 55
comparison ∼, 51
revised affine ∼, 58
method
Bauer–Skeel ∼, 110
Direct ∼, 107
Generalized Expansion ∼, 133
Hansen–Bliek–Rohn ∼, 113
metaheuristic ∼, 140
Parametric Direct ∼, 122
Parametric Interval Global Optimization ∼, 137
Parametric Linear Iterative ∼, 121
Parametric Quadratic Iterative ∼, 122
midpoint, 3
mignitude, 4
monotonicity approach, 118
multiplication
Chebyshev minimum-error ∼, 38
standard ∼, 36
trivial ∼, 35

N
noise symbol, 26
external ∼, 34
internal ∼, 33
norm
scaled maximum ∼, 53
normalization, 59

P
parameter
first class ∼, 90
zero class ∼, 90
parametric
hypersurface ∼, 90
linear programming, 165
parametric interval matrix
Hurwitz stable ∼, 78
inverse stable ∼, 81
positive (semi-)definite ∼, 74
Schur stable ∼, 79
pre-conditioning, 64, 68

R
radius, 3
applicability ∼, 170
regularity ∼, 59, 72
spectral ∼, 54
stability ∼, 60, 81
radius:strong regularity ∼, 73
range, 7
joint ∼, 27
regular
∼ interval matrix, 55
∼ parametric interval matrix, 60
rounding
directed ∼, 22
optimal ∼, 22
outward ∼, 22

S
singular
∼ parametric interval matrix, 60
slope, 12
solution
hull ∼, 99

inner estimate of ≈, 99
outer interval ∼, 99
parametric ∼, 100
solution set
AE-∼, 86
controllable ∼, 87
generalized ∼, 85
parametric ∼
AE-≈, 87
controllable ≈, 88
generalized ∼, 87
tolerable ≈, 88
united ≈, 88
tolerable ∼, 86
united ∼, 86
strongly regular
∼ interval matrix, 55

T
Taylor
∼ expansion, 14
∼ model, 14

W
width, 4

Printed in the United States
By Bookmasters